The Irish Battalion in
The Papal Army of 1860

The Irish Battalion in the Papal Army of 1860

By G. F.-H. BERKELEY

Edited by Brendan Cassell

Published by Papal Zouave International
2025

Copyright © Papal Zouave International 2025.

The Irish Battalion in the Papal Army of 1860 is a republication from Papal Zouave International of George Fitz-Hardinge Berkeley's work published by Talbot Press in Dublin and Cork in 1929. The editor has for the first time, republished the book in almost a century.

A special thanks to Christendom College's St. John the Evangelist Library for providing the editor with a copy of *The Irish Battalion in the Papal Army of 1860* and to Ave Maria University's Canizaro Library for providing high-quality scans of the book's illustrations.

For More information regarding this title and Papal Zouave International please visit:

PapalZouave.com

All rights reserved
ISBN: 979-8-9893230-4-3
Cover design by: Alexandr Chernushkin

Founded in 2023, Papal Zouave International is a historical society dedicated to promoting and preserving the memory of the Papal Zouaves. A unit of brave Catholic soldiers who came from across Christendom to defend the Papal States and Bl. Pope Pius IX during the 9th Crusade, between 1860–1870.

To learn more, visit PapalZouave.com

To My Wife
Caroline Isabel Berkeley

Editor's Preface

THE IRISH BATTALION OF THE PAPAL ARMY OF 1860 was authored by George Fitz-Hardinge Berkeley (January 29 1870–November 14 1955) and published in 1929[1] by Talbot Press in Dublin and Cork. Berkeley's work is unique in that it is the only book in English about the Irish Battalion where the author interviewed soldiers themselves, including one of their officers Through numerous letters, he corresponded with veterans and the families of those who had passed. To locate these veterans, he employed various methods, including newspaper advertisements and referrals from other veterans. He even hired individuals to track down and interview veterans on his behalf. His search extended beyond Ireland, reaching Irish immigrants in various countries. In America, he enlisted J.D. Hackett of 66 Jamaica Avenue, Flushing, New York, to locate and interview Irish Battalion veterans.[2] Dozens of these veterans had immigrated to the United States over the years, especially during the Civil War. Several went on to have notable careers in the U.S. Army such as General Daniel Keily, General John Coppinger, and Lieutenant Colonel Joseph O'Keeffe.

---

1   Ironically, the same year the Lateran Treaty was signed, thus ending the Roman Question.
2   *Correspondence of George Fitz-Hardinge Berkeley.* (National Library of Ireland: 1909-1939.)

Perhaps the most famous of the bunch is Captain Myles Keogh. Due to the Papal State's loss in the 1860 campaign and the subsequent deportation of most of its members, the Battalion of St. Patrick was reorganized as the Company of St. Patrick on November 9, 1860. remained in the Papal Army but found garrison life unfulfilling. In March, 1862 he and Keily left for the United States to join the Union Army. Keogh had a distinguished Civil War career, attaining the rank of lieutenant colonel. After the war, he commissioned in the regular army as a second lieutenant. A few months later he was promoted to captain and reassigned to the 7th Cavalry at Ft. Riley in Kansas under Lieutenant Colonel Custer. He met his fate at the infamous Battle of Little Bighorn on June 25, 1876, where he died leading I Company during Custer's last stand. The Sioux had a tradition of scalping the bodies of those they had slain, however, Custer's and Keogh's bodies were spared the mutilation. It is speculated that Custer may have been sparred due to his status as the regiment's commander. While Keogh may have been spared because he had on him the Pro Petri Sede medal,[3] which was awarded to Papal Soldiers who fought in the 1860 campaign. The Sioux were somewhat catechized, so they recognized the cross of St. Peter on the medal and spared his body.[4] Some say Sitting Bull took the medal and wore it himself.[5] Some of the Sioux warriors described Keogh as one of the "bravest" men they ever fought, and that he was the last soldier to fall.[6] The only survivor of the battle

---

3   The original medal, along with his Knight of the Order of St. Gregory the Great medal, was destroyed in a fire in 1865. He replaced them two years later. John P. Langellier, Kurt Hamilton Cox, and Brian C. Pohanka. *Myles Keogh: The Life and Legend of an "Irish Dragoon" in the Seventh Cavalry.* (El Segundo, CA: Upton and Sons, 1998), 163.
4   Some historians claim it was an Agnus Dei medal or his St. Gregory the Great medal. Ibid.
5   There's also some evidence to suggest the family could have retained the medals. Ibid.
6   Ibid., 137.

was Keogh's horse Commanche.

Many Irish Battalion veterans settled in New York and established the Papal Veteran Association of the United States, which grew to over 250 members across the country, with the majority based in New York.[7] Through open to other veterans of other papal units such as the Papal Zouaves, the association was primarily comprised of Irish Battalion veterans. They regularly participated in parades, masses, and funerals, often wearing their uniforms. Notable events include the 50th anniversary of Pope Leo XIII ordination.,[8] New York's centennial celebration,[9] and requiem Masses for Bl. Pope Pius IX.,[10] John McCloskey, (America's first cardinal),[11] and Pope Leo XIII.[12] Each year during the third week of September, the association held a reunion and requiem mass to pray for the repose of the souls of their fallen comrades and to commemorate the battles of Ancona, Spoleto, Perugia, and Castelfidardo during the 1860 campaign[13] and the siege of Rome that led to the fall of the Papal States on September 20, 1870.[14] On St. Patrick's Day they would participate in a parade and attend a lecture about the saint.[15]

One of the associations most significant moments occurred on February 1, 1880, when an award ceremony was held to distribute the Pro Petri Sede medal to Irish veterans who had not yet received one. Originally, the medal was to be awarded to all Papal soldiers who fought in the that campaign. However, most of the Irish soldiers were imprisoned and deported before the medals were issued on November 12 1860. Later efforts by

---

7   *New York Herald,* February 11, 1878.
8   *New York Tribune,* January 1, 1881.
9   *The Evening World,* May 1, 1889.
10   *The Baltimore Sun,* February 16, 1878.
11   *New York Times,* October 16, 1885.
12   *New York Tribune,* July 30. 1903.
13   *New York Tribune,* September 19, 1881.
14   *Utica Daily Observer,* 22 September 1879.
15   *New York Herald,* March 18, 1879.

the Papal government to distribute the medals in Ireland were met with challenges, some were even sent to the wrong people. As a result, many Irish Battalion veterans never received their medal.[16] Nearly two decades later many of the Irish veterans who had earned the medal were finally receiving them in the United States. The medals were presented by Louis Binsse, former lay consul-general of New York City. The rest of the event went as follows:

> Letters were read from General Kanzler, Commander-in-Chief of the Pontifical Army, Alfred La Roque, President of the Union Allet of Montreal, the veteran organization of the Canadian Papal Zouaves, Colonel Charles Tracey, of Albany,[17] and John D. Keilley, of New York. There were about thirty of the veterans present in uniform.
>
> After a few introductory remarks by the chairman, the choir of the Cecilia Society of St. Francis Xavier's Church sang the Pontifical hymn. The Rev. Michael Coughlin, pastor of Fausse Pointe, La.,[18] an ex-sergeant of the Battalion of St. Patrick, Irish Brigade, spoke of the pleasure it gave him as a soldier of the Papal Brigade of twenty years ago to take part in the proceedings.
>
> Father Preston said the evening should be a memorable one to all present. It was said to be sweet to die for one's country, but he thought it was sweeter to give one's life for the Vicar of

---

16   Nicholas Schofield, *Victorian Crusaders: British and Irish Volunteers in the Papal Army 1860-70*. (Warwick, England: Helion & Company, 2022) 77.
17   Tracey was a former Papal Zouave and fought under Captain D'Arcy during the Siege of Rome on September 20, 1870. He later served as a congressman for New York from 1887-1895.
18   Faussee Pointe is a region in Louisiana.

Christ. He expressed confidence in the ultimate success of the movement for the restoration of the Pope's temporal power. He characterized the statement that the majority of the inhabitants of the Papal States had desired the abolition of the Papal sovereignty as false and claimed that the ballot which was taken was fraudulent.

Medals were then presented to the following: Sergeant-Major Hynes, Sergeant Michael Ledwidge, Sergeant Richard Murphy, Sergeant Forrer Brereton, Michael Buckley, Edward O'Brien, John Delaney, Patrick Mulligan, John Quinn, Michael O'Donnell, William O'Brien, and Patrick Quille.[19]

The veterans of the association retained their militant spirit long after their service in the Papal Army. Perhaps the best reflection of their dedication to their faith and Pontiff is contained in a resolution they passed upon the death of Bl. Pius IX.:

> The members of the Association of Irish Veterans of the Pontifical Armies, in special meeting assembled, having heard of the death of their beloved sovereign, the late Holy Father, Pius IX, do hereby resolve:
> That, for his loss they feel a profound sorrow, tempered by the belief that his eminent virtues, his heroic defense of the liberties, temporal and spiritual, of the Church; his sublime submission in adversity to the will of Divine Providence, have earned for him a place among the elect at the right hand of his Master;
> That, standing around his bier, we pledge

---

19 *New York Tribune*, February 2, 1880.

> ourselves anew to the support of the principle which he advocated and so strenuously defended—the independence of the Supreme Pontiff as a temporal prince—and we hereby renew, for the maintenance of that principle, the offer of our lives to his successor, who, we feel confident, will be one to whom the destinies of the Church may be confided.[20]

The history of the Irish Battalion is also closely tied to the Papal Zouaves. As most of the members of the Company of St. Patrick who stayed in the Papal Army integrated with the Papal Zouaves after their unit was disbanded on October 1, 1862. Many went on to have notable careers in the Zouaves such as Captains Albert Delahoyde and James D'Arcy. Both of whom would go on lead companies composed of largely Irish and English-speaking Zouaves.

Through George Berkeley was a protestant, he provided a fair and accurate report on the history of the Irish Battalion and the Risorgimento. Unlike Patrick Keye's O'Clery, an Irish Papal Zouave who wrote two large volumes on the history on Italian Unification from a Catholic perspective.[21] Berkeley was motivated by his Irish heritage rather than his religious affiliation. He began writing *The Irish Battalion* in 1911 but was forced to stop due to the outbreak of World War I. After the war, he was delayed again as he had to move to Italy in 1920 due to health concerns with his wife.[22] However, this turned out to be a blessing in disguise as he was able to access more resources to write for his book. Following its publication in 1929, he began working on a three-volume work on the history of Italian Unification between 1815-1848. His series

---

20 *New York Herald*, February 11, 1878.
21 These works were republished as a single volume by Papal Zouave International in 2024.
22 It is unclear how this decision benefited her health.

Italy in the Making further demonstrates his commitment to historical accuracy on the Risorgimento when many previous authors were merely Piedmontese apologists. For example, he spent a significant amount of time interviewing Papal Army veterans and those from the opposing side: "In the course of thirty years I have met men of all parties—old Garibaldians, Piedmontese, Papal Zouaves and other survivors of the great days, and of course hundreds of a more modern date; and I have discussed every possible side of the question with them."[23] Additionally, he expressed his desire to only tell the truth: "The truth, the whole truth, and nothing but the truth—with the help of God."[24] And lastly, he sought to be fair: "In volume 1 the principal hero was Charles Albert: in the present volume it is Pius IX. The latter has never received the credit due to him for his splendid effort during his first two years."[25] In summary, Berkeley was a serious historian and his works have done much to preserve the history of the Irish Battalion.

This edition retains the original publication's style while making minor updates for modern readability. Changes include converting marginal notes into footnotes and standardizing spelling from British to American English (e.g., colour to color). We hope you enjoy The Irish Battalion of the Papal Army of 1860, republished for the first time in nearly a century!

---

23   G. F.-H. Berkeley, *The Making of Italy, Volume I*. (New York: Cambridge University Press, 1932), ix.
24   Inside cover of G. F.-H. Berkeley, *The Making of Italy, Volume II*. (New York: Cambridge University Press, 1936.)
25   Ibid., xv.

Eight members of the Papal Veterans Association of the United States act as an honor guard at the reqiuem Mass for Pope Leo XIII, in St. Patrick's Cathedral New York City, on July 30, 1903.

Pictured are Captain John Kirwin, John O'Connell, Sergeants Richard Murphy, P. C. Dooley (Papal Zouave), Edward Ledwidge, John E. Moriarty, James Murphy and Cornelius O'Leary.

From the *New-York Tribune,* July 30, 1903.

Author's Preface

THIS book is only a short record of an episode in our history, but it is written in fulfilment of promises made to the old survivors of the expedition of 1860, both in Ireland and in America. And we owe it to them that the episode should be remembered, for it represented much honest endeavor and self-sacrifice on the part of men who cared for Ireland at a time when politically she was, perhaps, at her lowest ebb. Moreover, from the general point of view, its subject is not without interest; it certainly throws a side-light on a very great period, that of the Italian Risorgimento. A proof of this fact is the reception accorded to our documents by the officers of the Italian War Office, for whose kindness and courtesy it would be impossible to make sufficient acknowledgment.

It is almost the last of the "Irish Brigades," that is to say, of the Irish corps serving under foreign governments; but this "brigade" differs essentially from the others, because it was enrolled for a particular cause which lay very near to the heart of the Irish people, namely, the defense of the Temporal Power.

At the same time, it is also an expression of Irish nationality in days when the nation had no flag, and it is one of the few which obtains a mention outside Ireland.

As my search was originally started in the year 1911, I was able to find still among us some fifteen to twenty of the old men who had "fought for the Pope" in 1860; they were dotted about all over Ireland in farms, cottages, or town-lodgings, and three or four of them were in workhouses—in reality the most interesting old men in the whole country, where, however, few people ever gave them a thought. And, no doubt, there were others to be found scattered here and there all over the world. Thus, for instance, I received several answers from America. But many of the "brigade" were dead; for the men of 1860 lived in a period of war, and—there being no Irish national army—scores of them had been killed fighting for foreign causes, more especially during the four terrible years of the American Civil War. However, I found that one could get invaluable assistance from conversations with the veterans still alive, nearly twenty in number, and also out of letters from relatives of those dead, of which I must have received over two hundred. These narratives were sifted and then verified during a search of months in the Royal Archives in Rome and in the Historical Section of the Italian War Office; also, as far as is permitted, in the Vatican Library and Archives. For permission to delve among these thousands of papers I owe my sincerest thanks to the British Ambassador in Rome; also, to Commendatore Casanova, Director of the State Archives, and Cavaliere Loewinson and Cavaliere Polidori, archivists there; to the late Colonel Vigevano and to Colonel Cesari, C.O's of the Historical Section of the Italian War Office; and to Monsignor Mercati, Prefect of the Vatican Library and to his brother Monsignor Mercati, Prefect of the Vatican Archives.

Of the books on the campaign there is a list at the end of this volume, but in the course of this broken search there must have been many others read whose names have since been lost.

Other friends to whom I am greatly indebted are the Marchese Degli Azzi Vitelleschi, the well-known author of *Le Stragi di Perugia*, of *La Liberazione di Perugia*, and other works, who was also for years the editor of the *Archivio Storico Risorgimento Umbro;* to Commendatore Menghini, Director of the Risorgimento Library in Rome; in Ireland, to Mr. William O'Reilly (son of Major O'Reilly, the defender of Spoleto) who gave me his father's diary; to Miss Edith O'Reilly, who has been kindness itself in replying to my enquiries; to Sir John O'Connell; to Dr. De la Hoyde, brother of the well-known Papal Zouave; to the Marquis McSwiney of Mashana-glass, who so generously handed over to me the result of his long researches, including the valuable documents obtained from his friend Baron Kanzler, son of the victor of Mentana; and, most of all, to Mr. G. J. Bergin (the son of one of the volunteers) whose father's house had for years been a meeting place for the veterans, and who was therefore able to open up for me their whole tradition as it had been handed down since 1860; and also to friends such as Miss Weld, who has given me an immense amount of time and work.

Among the survivors of the "Brigade" there was only one officer, namely, the late Mr. M. T. Crean, but he was the one of all others (Major O'Reilly being dead) whom one would have wished to meet, for he had been wounded during the defense of Spoleto in 1860, and had, in fact, been the subaltern in command of the gate which the Piedmontese tried in vain to take by storm. He had thus been the central figure at the most dramatic moment in the story of the expedition; while his men and the Piedmontese Bersaglieri were bayoneting one another between the loosened planks, he had received a bullet through the right biceps, at only three or four yards range, and for his services on this occasion was afterwards rewarded with the Cross of the Ordine Piano, very rarely granted to subalterns. On his return to Ireland he was called to the Bar, became a well-known barrister in Dublin, and towards the end of his life accepted a post as one of the legal advisers to the Land Commission. He had only one son who served in the Army

Medical Corps during the Boer War, and—"like father like son"—won a Victoria Cross.

Throughout his life Mr. Crean had been able to keep in touch with some of his brother officers, and I had many long and enthusiastic talks with him about the Irish volunteers of 1860, a subject in which his interest never failed.

There was another veteran of Spoleto, a private named Edward Dunne, of Dublin, who was first discovered and interviewed for me by Sir John O'Connell. He was an excellent little man, evidently a born soldier, matter-of-fact, sensible, and extraordinarily accurate in his recollections, with an obvious contempt for anything "high-falutin'." His clearness of memory was doubtless due to the fact that on his return home he had enlisted and served for ten years in the British army, thus maintaining his interest in soldiering, whereas some of the other veterans retained only a rather blurred vision of three "hectic" months of training, ending in a fight, some fifty years ago.

During the campaign there were four principal engagements at each of which an Irish unit was present, namely, those at Perugia, Spoleto, Castelfidardo, and the siege of Ancona. My first aim was to find men who had actually fought on these occasions.

For Spoleto, I was able, as already stated, to get excellent first-hand information from Mr. Crean and Edward Dunne; also from two Italians, Signor Santorelli, who as a boy had been one of the excited spectators watching the fighting and retained vivid recollections of it, also from an old gunner named Luigi Vallini, who had taken part in the defense, but had little memory left.

For the fight at Perugia there was a well-known veteran, Brother Aloysius Howlin of St. Patrick's Monastery at Mallow. He wrote me several long letters describing his experiences, and these are included among the documents; and in Dublin

I had a long conversation with a veteran named MacCorry, whose information is also included in the book of documents which I hope to publish.

For the battle of Castelfidardo it was impossible to find anyone. There had only been one reduced company of Irish present at that engagement, and none of the survivors were to be found. For the company at Castelfidardo, therefore, it has been necessary to rely almost entirely on documents and books. But I was able to get some second-hand information from an old Papal Zouave, Mr. W. P. Ryan, the well-known President of the Irish Literary Society. He had had a friend who fought at Castelfidardo.

For the siege of Ancona it was easy to find veterans, as there had been four companies, or over 450 Irishmen in the garrison. In Dublin my chief help came from Mr. Mackey, head of the well-known firm of builders, and Mr. Kenny; also the veterans O'Callaghan, Fallon, Hobbins, and Byrne. In Sligo there were three—Macalyn, Brian Finn, and Ward by name. Elsewhere there were others with whom I exchanged letters.

It has thus been possible to get first-hand evidence for three out of the four engagements, and the descriptions of the fighting at Perugia, at Spoleto, and at Ancona were each submitted for criticism to men who had themselves taken part in it.

The great misfortune during this search was that it was cut short at a critical moment: it was begun in 1911; then came the political strife of 1913 and 1914, and after that the Great War and subsequent troubles, creating for me a complete interruption of seven years, at the end of which the veterans were no longer to be found.

The list of books appended is comparatively short. » For any student of the Italian Risorgimento it would be easy to give a far longer list. But, primarily speaking, this small memoir is not compiled from books or newspapers. For the story of a military unit there can be only one reliable source of information, namely the documents of that unit—company, battalion, brigade, and army orders, reports, gazettes, letters, and other returns. Some thousands of these are preserved in the Archives, and I hope, perhaps, to publish a hundred specimens of them in a second volume, as they would be especial interest to Irish readers all over the world.

Contents

	PAGE
INTRODUCTION	1

PART I: THE PERIOD OF TRAINING BEFORE WAR BROKE OUT. (CHAPTERS I. TO VIII.)

CHAPTER I.—PUBLIC FEELING IN IRELAND	11
CHAPTER II.—THE STARTING OF THE VOLUNTEER MOVEMENT	21
CHAPTER III.—SETCH OF THE PAPAL ARMY	28
CHAPTER IV.—THE SECRET WORK OF THE REVOLUTION AND THE PRESS CAMPAIGN	41
CHAPTER V.—THE DIFFICULTIES OF THE FOUR SOUTHERN COMPANIES	51
CHAPTER VI.—THE FOUR COMPANIES IN ANCONA ...	68
CHAPTER VII.—MAJOR O'REILLY'S WORK AT SPOLETO	76
CHAPTER VIII.—IMPROVEMENT	84

Contents

PART II: THE CAMPAIGN OF CASTELFIDARDO, 1860.
STORY OF THE IRISH VOLUNTEERS DUING THE FIGHTING.
(CHAPTERS IX. TO XX.)

CHAPTER IX.—THE PIEDMONTESE PREPARATIONS FOR INVASION 97

CHAPTER X.—WAR DECLARED. THE STRATEGICAL PLANS ON EITHER SIDE 108

CHAPTER XI.—POSITION OF THE IRISH BATTALION AT OUTBREAK OF WAR 112

CHAPTER XII.—THE CAPTURE OF PERUGIA 114

CHAPTER XIII.—AFTER THE FALL OF PERUGIA ... 126

CHAPER XIV.—AN IMPASSE 133

CHAPTER XV.—LAST PREPARATIONS 141

CHAPTER XVI.—THE FIGHT AT SPOLETO, SEPTEMBER 17TH 156

CHAPTER XVII.—NO. 4 COMPANY'S MARCH TO CASTELFIDARDO 176

CHAPTER XVIII.—THE BATTLE OF CASTELFIDARDO, SEPTEMBER 18TH 180

CHAPTER XIX.—THE FOUR IRISH COMPANIES DURING THE SIEGE OF ANCONA 196

CHAPTER XX.—THE END 224

Contents

APPENDICES.—Appendix A. Cardinal de Mérode's farewell order to the Battallion of St. Patrick ... 233

Appendix B. Decorations won by the Battalion of St. Patrick 235

Appendix C. Officers of the Battalion of St. Patrick in 1860 241

Appendix D. Table of Data from the State Archives of Italy 248

Appendix E. The number of Irish who fought in this campaign 260

Appendix F. The Irish casualties 264

Appendix G. Notes on the losses at Spoleto 266

BIBLIOGRAPHY 269

THE IRISH BATTALION IN THE PAPAL ARMY OF 1860

INTRODUCTION.

As this memoir is not concerned with the history of Italy, nor even with that of a campaign, but is merely the record of a single volunteer battalion which only remained in being for about three months and many of whose men had not been with it for more than half that time, there is no reason to embark on a lengthy description of the political and military situation in 1860. In any case it has already been dealt with in great detail elsewhere.[26]

We need only recall the fact that in the years 1859 and 1860 the whole peninsula was still torn between two sets of ideals; on the one hand the old ideal of loyalty to the historic states and to the Church; on the other the enthusiastic desire to create a new nation, to set up an united and independent Italy. And they had been years of extraordinary interest; in 1859 Cavour had achieved the great aim of his life, when he brought Napoleon Ill's army across the Alps to break up the Austrian military machine for him on the glorious days of Magenta and Solferino. Then on July 11th, 1859, there followed the peace of Villafranca by which the Austrians retained only the province

---

26   The facts of this Introduction are drawn from the general Italian histories, of which that of Tivaroni seems to me perhaps the best.

of Venetia, while a fairly strong Italian state was formed in the north under Victor Emmanuel, King of Piedmont. This northern kingdom regarded itself from the very first as the nucleus of the new Italian nation. But further south all was unsettled; it still remained to be seen what the other small states would do.

At the very southern end of the peninsula was the King of Naples, who certainly had no intention of joining the national movement; nor—from motives of conscience—had Pope Pius IX in Rome. But between the territories of the Church and Piedmont there lay the central states of Tuscany, Modena, and Parma, whose attitude as yet was uncertain; they might wish to remain independent units, or they might prefer to form an Italian confederation; or they might vote for fusion with Piedmont.

The following is a very short summary of their progress during the succeeding year.

After the battle of Magenta, all Austrian protection being withdrawn, the sovereigns of Modena and Parma decided to leave their states; in Tuscany, the Grand Duke had already said good-bye to his capital. These departures tended to simplify the general situation, and by September 11th the three states had recorded a vote in favor of union with Piedmont. But what especially concerns our present subject is the fact that a portion of the Papal State was taking part in this movement and had decided to secede from the pontifical government in order to join Northern Italy at the very earliest opportunity; in fact as soon as the Austrian garrisons had marched away northwards, the Papal town of Bologna, with all the province of Romagna, had risen in rebellion and voted for a provisional government under King Victor Emmanuel, and very soon afterwards was being administered by a Piedmontese *commissario* or representative. This defection was a serious blow to the Roman government, and indeed soon proved to be the beginning of the end of the Papal State: but it can hardly have come as a surprise, for ever since the time of Napoleon I, these northern provinces had been in a condition of constant revolt against their inclusion in the Roman territory, and even at the Congress of Vienna in 1815 some efforts had been made to set them up as a separate entity. Their disloyalty was of old date and due to various causes; partly to national sentiment; partly to the fact that a modern democratic town such as Bologna would always hate a priestly government and would be certain to feel the commercial pull towards the northern business communities with which it had been connected in Napoleon's day; partly also to the old inter-municipal jealousy between Bologna and Rome. One may say, in fact, that the wisest course for the Pope might perhaps have been to let Romagna go in peace. But this, unfortunately, was a matter of conscience with Pius IX, just as it had been with previous Popes: he considered it his bounden duty to hand on the Papal State intact to his successor, because he regarded these territories as being conferred on him by a Higher Power for the welfare and independence of His Church.

Thus, before the end of the year 1859, the new north Italian realm had virtually extended its authority down to the southern frontiers of Tuscany and far into the Papal State. But it aimed, of course, at spreading still further and uniting the whole of Italy under one crown, and with this end in view had already made various efforts to raise rebellion within the Papal provinces of Umbria and the Marches so as to have a pretext for invading them. For the moment, its chief difficulty lay in the unsympathetic attitude of France. Napoleon III was at first hostile to these schemes; but throughout that autumn he was gradually coming to the conclusion that he could not prevent the march of events and consequently that his wisest course was to claim compensation for permitting them. Finally, in the spring, a bargain was struck: on March

24, 1860, Napoleon agreed to accept the provinces of Savoy and Nice, as his price for allowing Piedmont a free hand in Italy, and from that day onwards, he and Cavour became "accomplices"; this at all events was the word used by Cavour himself with reference to their next line of action. For the time being, of course, these arrangements were kept secret in order to avoid offending French Catholic opinion, but the trend of events was fairly apparent to the Pontifical government which began to perceive that it must take some steps in self-defense.

It is unnecessary here to deal with the long controversy which produced *Le Pape et le Congres* and other brochures. They were simply attempts to persuade the Pope to resign his claim to Romagna, and later on, his actual sovereignty of Umbria and the Marches, in return either for money or for some nominal vicariate which might save his face and perhaps salve his conscience. It was hoped thus to gather the Pope's state into the new nation without further bloodshed. But for Pius and his churchmen this was a genuine case of conscience: principle against principle; so the offers were refused and finally, when Romagna voted itself into northern Italy under the House of Savoy (March 11, 1860), a Bull of Excommunication was issued against all concerned, including Victor Emmanuel himself (March 26, 1860).

From the Papal point of view, matters were becoming very serious: within the previous eight months, the united-Italy movement had spread downwards from the north, absorbing Romagna on its way, and was thus half-encircling the remaining provinces of Umbria and the Marches, endeavoring by every possible means to arouse and support rebellion within them, in order to have an excuse for invasion. And, one may observe, this accusation is not denied now a days by historians, not even by those who, like the present writer, sympathize profoundly with the makers of modern Italy and believe in her future: even they now freely admit the somewhat cynical opportunism of the methods used against the Papal State in 1860. But they justify them on the ground that Pius had been obliged to rely almost entirely on the power of Austria to preserve order in his state during the previous ten years, and they point out that the Ecclesiastical Government had become an impracticable anachronism,[27] and that it was impossible to form the Italian nation as long as an international state like that of the Papacy lay right across the center of it.

Still fate was dealing hardly with Pius IX; with him above all men it was dealing very hardly, for in his younger days he had been a good Liberal and a good Italian, by far more in touch with the modern world than any previous Pope, or indeed than almost any living prince. During his first years he had thrown himself into the policy of reform; he had, in fact, been himself the initiator of the last phase of the national movement which was now about to deprive him of his territories. He had done as much as it was possible for a Pope to do and had remained the champion of Liberalism until his Liberal minister Rossi had been murdered and he himself compelled to flee from the Papal State in danger of his fife (November 1848). It was only then that he had abandoned the cause of the people, because it was impossible for him to abandon the Papacy.

In 1860 the crisis was approaching; now that the Pontifical boundaries were conterminous with those of Piedmont, it was supremely easy at any moment for roving bands, organized, armed, and assisted by the revolutionary committees, to dash across the frontier, rush through three or four of the small towns, tear down the Papal emblems, and hoist the tricolor and proclaim far and wide that the district was in revolt and had declared for an united Italy. Unless these tactics could be checked it was evident that very soon Pius would be discred-

---

27   Editor's Note: O'Clery and other pro-papal historians on the Risorgimento would push back on this point.

ited before all Europe, and would stand in danger of seeing his state snatched away from him by raiders and rebels without ever having exerted his full power to defend it.

It was at this stage of the story, and then only, that the Papal Government decided to issue an appeal to the Catholic world for men and arms; and during the first quarter of the year 1860 it began to send out its emissaries to all the nations concerned. In the existing circumstances Pius could hardly do anything else, as his own state was far too small[28] and too deeply divided in sentiment to hold its own without foreign assistance. But it must be remembered that this new army of 1860 was raised only for the purpose of keeping order within the state and of repelling raiders from outside; it was not intended to cope with regular forces. For protection in case of any regular war of invasion the Papacy trusted to the Powers of Europe, guarantors of the treaty of 1815.

In April 1860 recruits began, to pour into Rome from all the Catholic nations, 'and continued to stream across Europe all through the summer, an extraordinary medley consisting of men of every class and race. There were many Swiss, usually German by language, and many Bavarians, Poles, Czechs, and Germans from Austria; some French, of whom the hardy Breton contingent was perhaps the best; some sturdy Belgian peasants; and, last to arrive, but in the highest spirits, an Irish "brigade." The bulk of the Pope's army, however, was composed of Italians from his own state, a point which is often overlooked by his detractors; as is also the fact that they were all raised by voluntary enlistment because he refused ever to sanction conscription.

At this point, however, we must leave the situation in Italy, to examine the trend of Irish feeling during these years.

---

28   The Papal State numbered less than three million inhabitants, whereas the new North-Italian state must by now have been about eleven or twelve million strong; and in resources, training and organization the difference was far greater.

Introduction 7

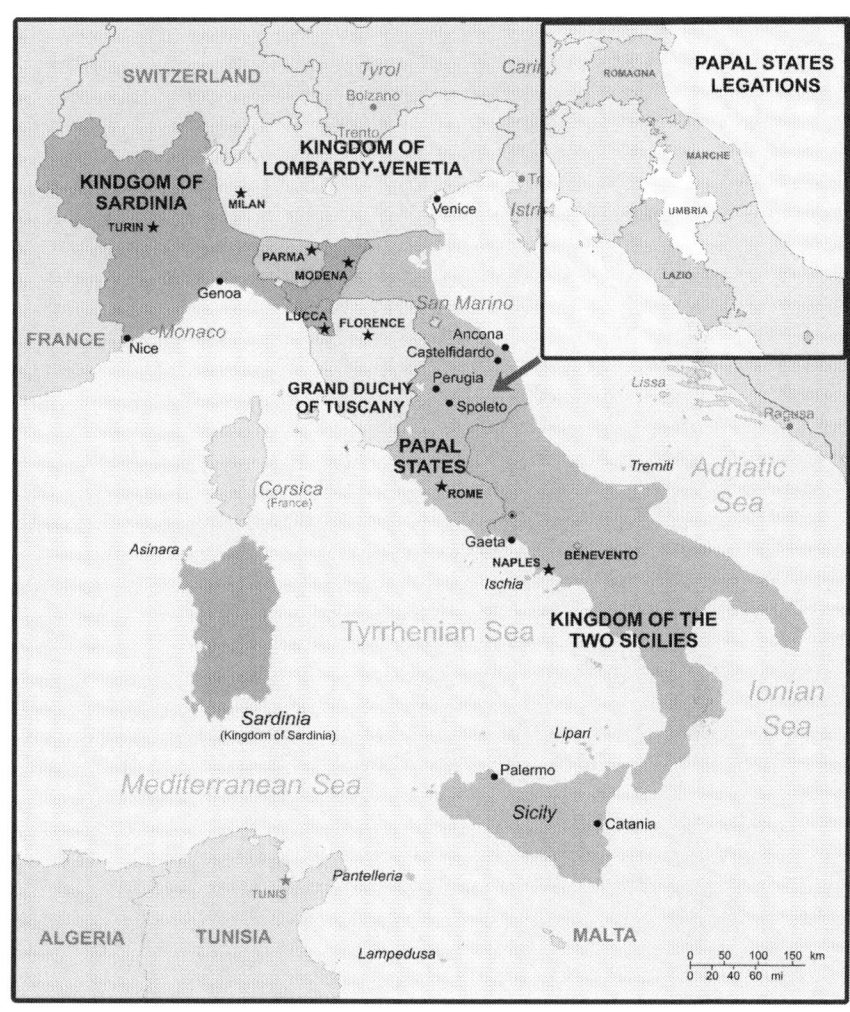

Map of the Italian Peninsula 1860.

Part I.—Chapters I. to VIII.
The Period of Training Before War Broke Out.

CHAPTER I.
Public Feeling in Ireland.

Italian students of the campaign of 1860 have always—and very naturally—regarded the Papal army as one single military unit; as one concrete whole, animated throughout by the same motives, that is to say either by religious enthusiasm or by desire for gain. On the one hand the Papal chroniclers can admit no other guiding sentiment at all among their ranks than that of pious self-sacrifice; on the other hand, historians of the Risorgimento, dismissing the religious motive as impossible, are led to regard the whole Papal force alike as merely a collection of ne'er-do-wells drawn to Italy solely by the need of pay or the hope of plunder. To the present writer, however, if he may be pardoned for saying so, each of these verdicts seems rather a superficial analysis of an exceedingly interesting phenomenon: in his belief the Papal army never reached the point of being one single military unit, but remained merely an agglomeration of separate corps drawn from the various Catholic nations—serving side by side, it is true but not always on the best terms with each other. The main though often unconscious motive that inspires each of them is in reality to be found in the history of its own native land: in the Austrian battalions, for instance, religious sentiment was certainly assisted and accompanied by the political ideas traditional to Austrian statesmanship; similarly the French Guides were undoubtedly fired by the old legitimist loyalty of France, that is to say, loyalty to their Church and to their King: in short, although the hopes of all were equally centered on the defense of Rome, each separate unit represented at heart the militant Catholicism of its own nation.

Of no corps is this so true as of the Irish battalion of St. Patrick: the motives that inspired it cannot possibly be understood except by a reference to history. Therefore, before attempting to describe it, we must begin by a very short review of the historical conditions of Ireland, going back as far as the eighteenth century and the Penal Code. This is regrettable, of course, because the Penal Code is a subject which we should all like to omit, and one would find it easier and far pleasanter to act upon the celebrated dictum of Sir Horace Plunkett about this being a subject for Englishmen to remember and Irishmen to forget. But—most unfortunately—the Penal Code is far and away the most important force that has ever exercised any influence upon the making of modern Ireland. No other is even comparable with it: it came at the turning-point in her history, at the critical moment when the old mediaeval life of the clans and colleges was dead and the new industrial and commercial development was about to take their place; it entirely distorted the growth of the new national life; and the modern Irish democracy, even to-day, bears the indelible, unmistakable impress left upon it by the laws of the eighteenth century. One may perhaps be allowed to suggest that there is one way, and surely one way only, of triumphantly requiting those iniquitous laws; and that is for the Catholics who now rule most of the island to prove, by the contrast of a perfect toleration, that they were always in the right, that the fears entertained of them were always groundless, and consequently that their century and a half of suffering can never possibly be justified.[29] And thus they will assuredly stamp on their race an ineffaceable hall-mark of nobility.

* * * * *

The Penal Laws reached their zenith of iniquity before the year 1780, but Catholic Emancipation was not won until the year 1829, and even after that there remained in being some remnants of the code, in fact down to the time about which we are writing. One of their principal results was to identify Catholicism in Ireland with nationality; for years to come the two terms were almost interchangeable; and so an attack on Catholicism—and this is important as concerning the subject of this Memoir—was regarded as almost equivalent to an attack on nationality. A weakening of the Catholic Church would have seemed to most Irishmen to be a weakening of the institution from which his nation drew a large share of its moral strength.

Moreover there was the sentiment of gratitude a thousand times deserved. Throughout the whole of the penal century it had been the priests of the Roman Church, often Jesuits, who had carried their lives in their hands daily while they toiled to preserve the individuality and culture of the Irish race. The result of that self-sacrifice had been that the Catholic religion had wound itself around the very heart-strings of the people; to them it had become the creed of the oppressed, the Mater Dolorosa, Our Lady of Suffering, in whose service every true man should be ready to live and die.

* * * * *

But there was another side to the history of the eighteenth century.

Every reader knows that all these years of misery at home are illumined by one constant ray of glory—the arms of her

---

29   It will be remembered that Mr. Froude, for instance, tried to justify the oppression of the Catholics in Ireland by saying: "You cannot tolerate those who will not tolerate you."

soldiers abroad; the "Irish Brigades" will never be forgotten.[30] In the archives and libraries of southern Europe there are thousands of documents lying unread and uncared for; documents in which their figure, half incongruously, the ancient Gaelic names of scores of officers whose whole life was an honor to their native land. That story is the epic of an oppressed race, though as yet very little of it has been written. And this is a pity, because to a superficial student of Irish history the years between 1690 and 1791 sometimes seem a period of unvirile weakness and submission.

Here one can merely summarize a few of the best known names connected with the tradition of the Irish in foreign service during the eighteenth century. In Spain we have the records of numerous Irish battalions which fought in the Spanish wars, and the best known name is O'Donnell. In Austria the names of O'Donnell and Taaffe, Nugent, and other noble families remind us of the Irish officers in the imperial service, the greatest man among them all being probably Marshal Browne, celebrated for having defeated Frederick the Great;[31] and remnants of the Irish tradition in Austria lasted

---

[30] Every reader has heard of them; but few realize that the numbers of Irish abroad must at times have amounted to a kind of floating population of soldiers. And these men, who in their own country were only a lower stratum, abroad were often entrusted with the most responsible posts, and were received as equals by the highest in the land.

[31] As an instance of this:—Col. Cavanagh has published an exceedingly interesting article in the *Journal of the Royal Society of Antiquaries,* for September 14th, 1926, giving the names of those Irishmen who became Knights of the Imperial Military Order of Maria Teresa. This order was founded in 1757, as a reward for officers distinguished in war; and during the next fifty years no less than 25 Irishmen appear on its roll. Theirs is a splendid record, one which would reflect glory on any nation, not merely for the many scores of campaigns in which they won honor, but also for the high commands and responsible posts which were often conferred upon them. Lack of space prevents one's doing more than mentioning their names.

    Lieut.-General Baron Thomas Plunkett.
    General of Artillery Count William O'Kelly.
    Lieut.-General Baron Francis Nangle.
    Colonel Chevalier Hume Caldwell.
    Field-Marshal Count Francis Maurice Lacy.

into the nineteenth century, in fact, almost down to our own time: in 1848 an O'Donnell was Governor of Milan during the Revolution; in the same year Marshal Nugent, one of the best Austrian generals of his times, won great credit by his leadership of the column which forced the passage of the Piave and brought reinforcements to Radetzky just before the battle of Custozza; in the following year he was mortally wounded at the siege of Brescia, and—a chivalrous touch—when dying he bequeathed half his property to that town in token of his admiration for her gallant defense.[32] But the tradition of Irish descent has perhaps been better maintained by the Taaffe family than by any other; they had a battalion there; and to this day the Austrian branch and the Irish branch still keep up their relationship.[33]

    Lieut.-General Count Philip George Browne.
    General of Artillery Count John Sigismund Maguire.
    General of Artillery Count Richard D'Alton.
    Major-General Baron John Baptist Purcell.
    Lieut.-General Count John O'Donnell.
    General of Cavalry Count Charles Claude O'Donnell.
    Lieut.-General James Robert Nugent.
    Major-General Baron John O'Brien, Count Thomond.
    Lieut.-General Count Patrick Oliver Wallis (Walsh).
    Major-General Count Henry O'Donnell.
    Lieut.-General Count William Lacy.
    Lieut.-Colonel James Patrick O'Mulrian.
    Colonel Baron James Bernard MacBrady.
    General of Artillery Baron Thomas Brady.
    Colonel Count William Mahony.
    General of Artillery Count John George Browne
        (son of the great Marshal Browne).
    Major Chevalier John Wilson.
    General of Cavalry Count Andrew O'Reilly.
    Field-Marshal Count Laval Nugent.
    Major Chevalier Peter Martyn.

    These names are those of the distinguished Irishmen in one order of knighthood alone. There must have been many others. One notices, for instance, that some of the most important names —Taaffe, for instance—do not appear.

32   Brescia, "the lioness of Italy," she is called by the poet: Carducci, had made a glorious defense, even when all was hopeless.

33   The late Mr. George Taaffe, of Smarmore Castle, Co. Louth, had in his possession a book of old family documents some of which related to their battalion in Austria; among them he had one letter of congratulation on his marriage

In Naples during the eighteenth century there was for many years an Irish battalion;[34] the archives there are a mine of interest which has not yet been fully explored. In Russia we find no brigades nor battalions, but Irish and Norman-Irish names—as for instance, de Lacey—among the naval and military officers from the days of Peter the Great onwards. In Germany there are said to be innumerable records as. yet unstudied. In the United States, so we have all read, during the War of Independence, some of George Washington's battalions "spoke almost as much Irish as English," these were the forerunners of the gallant "69th" of the following century. The French service, as everyone knows, was never without an Irish corps from the Treaty of Limerick in 1689 to the fall of Napoleon in 1815. I do not dwell nor enlarge upon the exploits of these heroes because they are already well known.[35] My only object in mentioning them is to show how, long before the

---

received from Count Taaffe, head of the Austrian branch, and addressing him in the second person singular—thee and thou—in token of relationship.

34   The Marquis MacSwiney, of Mashanaglass, has shown me scores of extremely interesting documents that he has found' there. He is, as everyone knows, engaged in publishing the results obtained by him in France and Spain as well as in Italy. *V.* his exceedingly interesting paper read to the Royal Irish Academy in January, 1927: "Notes on some Irish regiments in Spain and Naples in the 18th Century."

35   O'Callaghan gives the following summary of the number of Irish corps in the service of France between the years 1690 and 1791. ". . . . it appears that there were in French pay, from 1690 to 1692, 3 Regiments of Infantry; from 1692 to 1698, including Mountcashel's Brigade, King James's army from Limerick, etc., and counting Dismounted Dragoons as Infantry, there were 13 Regiments of Infantry (in 25 or 26 battalions), and 3 Independent Companies, 2 Regiments of Horse, and 2 Troops of Horse Guards; from 1698 to about the middle of 1699, there were 7 Regiments of Infantry, and 1 Regiment of Cavalry; and from the remainder of that year to 1714, 8 Regiments of Infantry, and 1 Regiment of Cavalry; from 1714 to 1744, 5 Regiments of Infantry and 1 Regiment of Cavalry; from 1744 to 1762, 6 Regiments of Infantry and 1 Regiment of Cavalry; from 1762 to 1775, 5 Regiments of Infantry; and from 1775 to 1791 3 Regiments of Infantry. The existence of so considerable an Irish force in France for a century after the Treaty of Limerick, proceeded, at first, from the attachment of the mass of the Irish people, as Catholics, to the representative of the Stuart dynasty, as deriving his origin from the old monarchs of Erin, as also a Catholic, as excluded on *that* account, from the Crown. . . ."—*History of the Irish Brigades in the Service of France.* Book III., p. 157. J. C. O'Callaghan.

period with which this work will deal, service under a foreign flag had become traditional in Ireland, where its glorious memories still fire the imagination of the people, even down to our own time.

*   *   *   *   *

For our present story the sequence of historical events is as follows: the Penal Laws may be said to have virtually come to an end when in 1829 the Catholics won the long struggle for emancipation.

In 1847 came that most appalling disaster of the nineteenth century, namely, the Famine. And during those terrible years Pope Pius IX—one of the kindest men in the world—earned heartfelt gratitude in Ireland. His efforts to help the starving peasantry have been described in detail elsewhere, so we need only summarize them. He started a subscription to supply relief-work and sent one thousand scudi out of his. own privy purse. He ordered a solemn Triduo to be held—three days during which sermons were to be preached and public prayers offered up; and, in an encyclical of March 15, 1847, he "aroused the charitable zeal of the bishops and of the faithful, to assist the Irish who were tormented by hunger and worn out by pestilence." Under Pius's auspices the great Liberal preacher Padre Ventura deployed all his burning Sicilian eloquence in helping their fund; on Sunday, January 24, 1847, he preached a passionate sermon in the Church of Sant' Andrea della Valle, describing the past suffering of Ireland and the present horrors of the Famine so vividly as to arouse a wave of sympathetic enthusiasm.

"Could it be said that by the sufferings of this famine this people is atoning for some sin unknown to us but inscribed before the tribunal of Divine Justice? No; a people of Christian heroes such as. that of Ireland cannot be accused of unknown sins. When we see the Innocent, the Uncontaminated, the Just, the Son of the Most High dying on the Cross to save sinners we are abundantly taught that the tribulation and suffering of a virtuous man may deserve benefits and grace from Heaven both for him and for others."

Only three months later came the tragic news that Daniel O'Connell had been unable to carry out his. last great hope of reaching Rome before he died; he had passed away in Genoa but had directed that his. heart should be taken to the City of the Church for burial. A great Requiem service was held in his. memory and Padre Ventura preached the funeral oration.[36]

These are only a few of the kindly thoughts of Pius IX towards Ireland; and it was natural that in 1860 the Irish people should wish to defend the benefactor who had helped them in the years of the Famine.

* * * * *

In 1859 the war of Italian Liberation broke out. It was obviously destined to appeal forcibly to the main elements of enthusiasm among the Irish; to their tradition of military glory abroad, and to their intense reverence and gratitude to the Papacy. It would have been impossible for them to watch unmoved a struggle between the French and Austrian armies, in each of which Irish officers were still to be found, or, later on, to acquiesce without resentment in the invasion of the Papal states. It was certainly the very moment for another "Irish Brigade."

At first public opinion in Ireland was mainly in sympathy with the cause of Italian liberation. The French were marching into Italy to free a subject population; and here and there throughout our island there still remained veteran survivors of the wars of Napoleon, old men in whose ears the Marseillaise sounded as a call to arms. So the Irish followed with keen interest and even with emotion the advance of that valiant army in which their compatriots had served through every

---

[36] V. The *Contemporaneo* for January 30th and April 17th, 1847 (the principal Roman newspaper of that date).

vicissitude of fortune during nearly two centuries; all the more enthusiastically because it was now under the command of one of their own race, General Patrick Macmahon, whose Irish name and features bore living testimony to the fact that he was the great grandson of one of the hardy patriots who had preferred exile to servitude after the Treaty of Limerick. And when, on the glorious day of Magenta Macmahon won his marshal's baton, there was great and widespread rejoicing throughout Ireland. Bonfires blazed on the hills of Clare, the ancient home of his family; a proposal that our people should present him with a sword of honor evoked a splendid response and was rapidly carried into effect; on his return to France he received, with the marked approval of the Emperor himself, an Irish deputation at Chalons, and after accepting the sword in his own name and in that of his son replied to their address in the most graceful terms, referring to Ireland as "the noble home of his ancestors." Certainly until the armistice of Villafranca the majority of our sympathies followed the Napoleonic arms.[37]

But while the Irish undoubtedly sympathized with the revolt of Italy against her Austrian masters, at the same time they were strongly opposed to any attack on the Papacy. That the Italians should rise for the sake of their freedom they fully understood, but in their eyes the Pope, himself an Italian, had quite as good, indeed far better claims to respect than Victor Emmanuel. Throughout two tragic centuries of history the ministers of the Catholic religion had been the devoted servitors of Ireland, the preservers of her traditions and of her nationality, indeed her unchanging friends when she had no others. Now was the moment to show gratitude. The policy of annexation was being joyfully preached by the

---

37   The priests, however, and those who followed their opinion were never much satisfied at the French invasion. They foresaw that it would endanger the existence of the Papal State.

most bigoted of the Protestant[38] newspapers; it was time to act. To the educated Irish Catholic in 1860 an attack made on the Papacy represented not merely a blow struck against the institution which he most revered, but also a change which he feared might indirectly affect his own country by weakening one of the sources of its moral strength. As to the internal affairs of Italy, he had scarcely better means of understanding them than the average Italian had of understanding the local politics of Ireland.

It was while public opinion was in this inflammable condition, that the decisive event occurred which led to the starting of this expedition.

---

[38] This, of course, does not mean that all Protestants were bigoted. Many of them believed that it was wisest for Italy to be unified, and that the Papal system of government was no longer possible. But unfortunately there existed also a widespread anti-Catholic hatred of the Pope, and for political reasons there was bitter resentment over the project of forming a distinctively Irish "brigade"; in England a Garibaldian battalion was being raised.

CHAPTER II.
THE STARTING OF THE VOLUNTEER MOVEMENT.

THE origin of the Irish Papal battalion might almost be called dramatic. One morning in March 1860, while Mr. A. M. Sullivan, the well-known editor of the *Nation,* was sitting in his office in Lower Abbey Street, Dublin, he received word that two visitors were waiting to see him. One of them, a friend of his, introduced the other, a stranger, as Count Charles McDonnell. He describes their interview in the following words:[39]

One day early in March, 1860, two gentlemen entered my office in Lower Abbey Street, Dublin. One was a friend whom I knew to be deeply interested in the now critical affairs of the pontifical government; the other was a stranger, apparently a foreigner. "Here," said my friend, "is a gentleman who shares some of those views you have been so strongly urging about defending Rome." I found in my unknown visitor Count Charles MacDonnell of Vienna, trusted attaché of Field-Marshal Count Nugent, and a chamberlain of the Holy Father. If ever chivalrous devotion to a fallen cause was personified, it was in this loyal and brave-hearted gentleman. He reminded me of those Highland chieftains whose attachment to the

---
39   *New Ireland,* p. 211.

Stuarts, romantic and tragical, evokes sympathy and admiration in every generous breast. Had he lived in the thirteenth century, he would have been a crusader knight; in 1641 he would have been a Cavalier; in 1745 he would have been at the side of Prince Charles Edward on the fatal field of Culloden. He came to see what Ireland would do—what aid she would contribute to the military defense of the Roman patrimony. "We know in Rome," said he, "that Garibaldi with the connivance and secret assistance of the Turin Government is organizing an aggressive expedition, but whether to strike at Naples or at us in the first instance we cannot tell. In any case we shall be attacked this summer. What will Ireland do for us?"

> "In the improbable event of the Government allowing volunteering, as in the case of Donna Maria," answered, "you can have thirty thousand men; if, as is most likely, they give no permission, but no active opposition, you will probably get ten thousand; if they actively prevent, nothing can be done. In my opinion, unless the proceeding is too glaring and open, Lord Palmerston will not raise a conflict, in view of Lord Ellenborough's letter and the 'million of muskets' movement on the other side in England. But the chief difficulty will be our own bishops. They will be adverse or neutral. Not one of them believes the little army of LaMoricière can cope with the overpowering odds of Sardinia."

The Count pulled from his breast a scarlet morocco leather case, and in five minutes satisfied me that abundant evidence had been secretly given at Rome by some of the crowned heads of Europe that if Monsignor de Mérode could, without French or Austrian intervention, defeat invasion by Garibaldian irregulars, Sardinia could be prevented from attacking.

This threw a new light on the situation. I think I can assert that it was on the faith of those private assurances the whole of General LaMoricière's movements were planned in 1860.

Out of this interview originated the idea of an Irish corps. And soon afterwards a very effective stimulus for recruiting was afforded by the publication of some sarcastic articles in one of the hostile newspapers to the effect that if the Pope could live on mere expressions of sympathy he would receive enough from Ireland to keep him alive for ever. This taunt aroused the people to action: "£80,000 and an Irish Brigade" was the cry raised in answer throughout the whole island. Monster meetings were held in various counties. The necessary organization having been rapidly discussed and decided on, it was not long before practical results ensued. Within a month the first batch of volunteers had left Ireland.

Thus, at only three or four weeks' notice, without any preliminary training and with only very little organization, was raised the Irish contingent for the Papal army. It consisted of men of all types, but, on an average, of a far higher class than was to be found in an ordinary British regiment: some peasants from the fields, some farmers, clerks, medical students, lawyers; in Dublin a big linen-draper's shop was suddenly emptied of assistants; some old soldiers, veterans of the Crimea or Indian Mutiny; some militiamen; and some Royal Irish Constabulary who had resigned their twenty to thirty shillings a week, with right to pension, for the pleasure of serving the Pope at twopence-halfpenny per day and their keep. A batch of young giants from Tipperary, a contingent of Gaelic speakers, mountaineers from Kerry, hardy, active, splendid material—but including rather unfortunately a rowdy gang of about fifteen boys who had been sent out of their parish by their priest because he wanted to be rid of them. This last-named batch, which was known as "The Kerry Boys" (although there were plenty of less troublesome Kerry boys as well) seems to have been rather unpopular among their comrades, and owing to its lack of discipline, tended, at first, to bring discredit on the battalion.[40]

It was manifest, however, that a movement of this nature would very soon encounter opposition. The Protestant party in Ireland did not like it; the pro-Italian party in England did not like it; and, from this and other points of view, the British Government was very unwilling to allow Ireland to form an Irish battalion for service on the opposite side. Moreover, Cavour complained about the enlistment in Dublin; so on May 16 a proclamation was issued from. Dublin Castle reminding all persons concerned that under the Foreign Enlistment Act any man entering the service of a foreign government was guilty of misdemeanor punishable by fine and imprisonment, as was also anyone helping him to do so; and that any master of a ship who knowingly conveyed such persons was liable to a fine of £50. This proclamation may very likely have lessened the numbers of the recruits, but it by no means stopped recruiting; throughout the whole summer they continued to embark, and it was difficult for the government to interfere, as there were some Garibaldians also going from England.

Towards the end of May and the beginning of June 1860, the various contingents of Irish boys began to arrive in Italy and continued to do so until almost the end of July. Some had come via Hull, Antwerp, Vienna, and Trieste: others who arrived later, came by Marseilles and Civita Vecchia direct to Rome. They were soon formed into a battalion of eight

---

40 No. 3 Company of the battalion was often called "the Kerry company," because there were about 30 Kerry men in it.

companies, at first over 1,300 strong, but afterwards reduced to under 1,100, called the Battalion of St. Patrick, though commonly known at home as "the Irish Brigade." It was not in reality a brigade; but, as four companies were kept in the northern garrison town of Ancona whereas the other four spent most of their time further south in the city of Spoleto, it was in fact divided into two entirely separate units, each of which possessed almost the numerical strength of a battalion. During their stay in Italy they saw nothing of each other, and even their methods of drill were not identical, the companies in Ancona being trained on the English system and those at Spoleto on that of the French army. They might, in fact, almost be considered as two separate corps, though both under the same commander, Major O'Reilly.

The selection of a commanding officer apparently caused some difficulty. The post had first been destined for another Irishman, Major Fitzgerald, a cavalry officer on the Austrian staff, and Major Fitzgerald afterwards arrived at Ancona and took command of the four Irish companies which were there. But owing to his captaincy in the Austrian army it was thought inadvisable that he should accept a battalion in the Papal service, complaints being already current that the Austrians were secretly arousing the Roman government against the Piedmontese. The choice therefore fell, most fortunately, on Major O'Reilly, a country gentleman well known in Ireland for his honorable and disinterested character. For him the undertaking, so unpopular with his own class, involved an exceptional sacrifice, owing to his position, and also owing to the fact that he was newly married; moreover, he was not a professional soldier—though he had been for years a keen student of military matters, and as a militia officer had obtained some experience of managing men. Yet in spite of these difficulties he threw himself heart and soul into the enterprise of which he soon proved himself a most capable guide: indeed there were moments when, but for him, it would have fared badly.[41]

When talking to the old men about their experiences one used often to be rather touched to find that the deepest impression on the minds of many of these country lads was made, not by the dangers of the campaign but by their first departure from Ireland.

Many of the veterans seemed still to have a singularly clear recollection of those first days of journey out to Italy; and when one remembers that in 1860 railways and steamers were more or less in their infancy, it is easy to realize what a deep impression would be left on the volunteers, especially upon those from the country districts of Ireland, by the absorbing interest of their first long voyage. Superimposed upon their inward uncertainties there soon came a feeling of freedom, of youth, and of adventure to come, all the more lively for their being suddenly plunged into the fife of the foreign towns where the people, the sunshine, the language, the uniforms, the bands, the cafes, and everything else were strange sights and full of enthralling novelty. And at every halt they were

---

[41]   Major O'Reilly was a country gentleman of the genuine type: a man who knew how to direct his own farming operations, who was a keen breeder of prize cattle, who rode to hounds, and in his younger days had kept half a dozen race-horses. Unlike the majority of his kind he was a patriotic Irishman; and he had the courage of his convictions; these facts alone would have accounted for his popularity with the people. In Italy the character which he earned was that of being a strict disciplinarian.

The following description of her father has been sent to me, most kindly, by Miss Edith O'Reilly:

"Medium height, dapper (his one little vanity being his small, well-shaped hands) a very stem face when he was thinking deeply, or explaining—for he was never anything but thorough, and he was short-sighted; but his face lit up with a very kindly smile when talking to friends or when amused. Very stern in the matter of justice, he was loved by the poor for the even-handed way in which he distributed it, even when they were sufferers."

The most valuable testimony, however, is always that which comes from the outsider. Mr. George Macaulay Trevelyan (whose famous books on the Risorgimento are familiar to us all) has told me that he still possesses a letter from his father recalling with pleasure "the honest weather-beaten face of O'Reilly," and referring to his popularity in the House of Commons, of which he became a member after his return from Italy. Mrs. O'Reilly had been a Miss Stafford-Jerningham.

received with the greatest kindness. They were the boys who were going to defend the Pope. Under these influences their natural high spirits soon reasserted themselves, and there are many stories which prove that before much time had flown the volunteers began to enjoy the life around them. During a day or two spent in Belgium we hear of cheery meals at restaurants and of many other "divarshuns," and of dancing at the cafes in the evening. But they had little leisure for becoming acclimatized; their stay at each place was short; sometimes it was only a question of hours before the whole batch would be whirled off in a train going south.

Of their journey there is nothing worth recording except one small anecdote which is rather pleasing. It tells of the arrival at Vienna of a batch of Gaelic speaking Kerrymen, and describes how, when their train ran into that great station, they stepped out on to the platform, shy and bewildered by the entire novelty of their position, feeling themselves strangers in a far-off country, uncertain as to what they should do, and without the slightest hope of making themselves understood. They were naturally rather alarmed, and—to make matters worse—not far from them they perceived a tall, athletic-looking officer whom, from the magnificence of his white Austrian uniform, they recognized to be a person of considerable importance, and therefore regarded with corresponding awe, unwilling to attract any kind of notice or criticism on his part. While they were still hesitating, this resplendent figure turned towards them, and then from his lips there suddenly fell the soft accents of their own native tongue and even of their own Kerry dialect of the mountains: "Sé bhur mbeatha, a bhuachaillí!" And what a pleasant greeting that must have seemed to these lost country boys! The staff officer turned out to be Major Fitzgerald, who had come down to meet them. He was a fluent speaker of Gaelic, having picked it up in childhood and never allowed himself to forget it.

CHAPTER III.

SKETCH OF THE PAPAL ARMY.[42]

AT this point it may be as well to give a very brief sketch of the army—or rather of the army in course of formation—in which the volunteers from Ireland were about to enlist. It was certainly the most curious in Europe.

The Papal State, consisting, before the loss of Romagna, of about three million inhabitants, was an entirely unwarlike institution. In reality a large estate restored to the Pope in 1815 for the maintenance and independence of the Roman Catholic Church, it was debarred from attacking any of its neighbors, and, for its defense, relied mainly on the guarantee of the great powers; and it was ruled by ministers of the religion of peace. The result was that the fighting spirit had almost died out, especially among the country people, who usually form the hardiest and best element for war. No man could be more peaceable and unpolitical than the peasant of Umbria; he was perfectly satisfied to till his small plot with its corn and vines and olives, and to live, marry, and die in the same way as his father and grandfather before him. In towns, on the other hand—with the exception of Rome and

---

42 This sketch is drawn mainly from the thousands of Papal military documents now lying in the Italian State Archives, also from those at the Italian War Office; from private Memoirs, and from the official histories of the campaign.

just a few others—there was a strong and growing anti-Papal element, sometimes rather turbulent, but usually very unmilitary owing to lack of training, and not very keen to volunteer. In no national revolution is there at first more than a small number of men who are actually ready to give their lives for the cause, and this was certainly true of these Umbrian town populations. Only a few of the citizens had yet reached the psychic condition necessary for rebellion, the state of mind when thousands would rather die than see their movement die; the majority were ready to agitate, to work, and to take the lesser risks, but not to go out into the field.[43]

As soon as Pius IX had realized that the governments of the great powers were no longer willing or no longer able to support him, he had decided to make an appeal to the Catholic population of Europe for men, arms, and money to defend him against the revolution. To carry out this more or less military measure it was necessary to have an efficient minister of war, and—Cardinal Antonelli being mainly a financier—the Pope's choice fell upon a comparatively young man, Cardinal de Mérode.

De Mérode, then, represents the military policy, for which he was far better suited than most churchmen: he came of an ancient noble family of Belgium, and consequently was in touch with many of the best known Catholics in that country and in France, and he had himself been a soldier; he had served for a year or two in the French Foreign Legion in Algeria.[44]

He at once threw himself enthusiastically into the work of raising a military force. It was an almost impossible task, but at the end of a few weeks he had achieved at all events one

---

43   *V.* Pimodan's letter, May 18—A.S.R Com.Gen. tr. pont. 3151.
44   He seems to have retained much of his military disposition and even appearance. I remember Mr. Crean (who had been a subaltern in the Irish battalion) telling me of the time when he was at Terni, how, on a beautiful summer evening he sighted some horsemen approaching at a smart canter. When they came near, to his astonishment he saw that their leader was Cardinal de Mérode, and that he was sitting his horse far more like a soldier than like a dignitary of the Church.

great success in securing the services of a friend of his early soldiering days in Africa, namely, General de La Morcière.

Nothing is more surprising than the very slight recollection that now remains of General de La Morcière, a man who at that time was considered the ablest soldier that France had produced since the days of Napoleon; but this is due to the fact that his career was cut short untimely; he is, in fact, a striking example of the wastage that occurs in national force owing to internal divisions and revolutions. Had he been born in England, his star might almost have outshone that of Kitchener in our own time; for, like Kitchener, he had begun life as an engineer officer; like him also he had proved his ability as the head of an intelligence department and as a military and civil organizer; he had also to his credit many well-known stories of his personal courage; and he had given infinitely greater proofs than Kitchener of his capacity as a leader on the field of battle. Africa is the grave of reputations; yet during twenty years of constant and difficult warfare in Algeria—when failures were almost as plentiful as during the Boer War—against de La Morcière alone no single error could be quoted. At the age of thirty-eight he was a lieutenant-general; on his return to France he was appointed a member of the commission for reorganizing the French army; in 1849 he was offered, and refused, the command of the Piedmontese army against Austria; under the French Republic he became Minister for War; but when the Republic was overthrown by Napoleon III de La Morcière went into exile rather than serve him.[45]

This was the man now secured by de Mérode. No better could have been found: in addition to all his other gifts he possessed a special talent for training young troops; he was one of the original organizers of the Zouaves, and was celebrated for having produced one of the finest fighting corps

---

45   Keller.

of the French colonial army out of a mob of Parisian riff-raff shipped out to him to get rid of them.

His chief of staff was the Marquis Georges de Pimodan, ex-colonel in the Austrian cavalry, the son of a French legitimist noble who had emigrated to Austria after the revolution of 1830 in France; he had been one of the most distinguished soldiers of his rank and age in the Austrian service, a colonel at the age of thirty two, and already a man of legendary courage.[46] But he had decided to return to his beloved France—fortunately for himself—some years before the outbreak of war in 1859. In 1860, though a married man with a family, he offered himself for the defense of the Papacy, but seems soon to have realized that the situation was hopeless. He is described as one who worked with unflagging enthusiasm but with evident sadness; a tall, quiet, silent man, of very courteous manners, and in military matters a strict disciplinarian.

With a fine sense of mutual understanding de La Moricière and his officers toiled all day and throughout a good deal of the night striving desperately against the thousands of difficulties which inevitably arise from lack of time and lack of money. There was almost everything to do: the existing Papal army consisted of twelve battalions of infantry, averaging about 800 strong, and one squadron of cavalry; but it was in an almost unimaginable state of disorganization; one need only say that some of its guns were found in a cellar. It was fortunate for de La Moricière that he understood how to do almost everything himself, from repairing a field gun to fortifying Ancona harbor, because at first there was no one else to help him and there were many to thwart him; if he had been allowed two or three years, he might possibly have established a working organization in spite of the numerous ill-wishers around him,

---

46   The story is well known of Pimodan's narrow escape from death, while a prisoner with the Hungarians. Having tried to escape, he was sentenced to death, and had actually written his last message to his family on the window-pane with his diamond ring, when an Austrian force recaptured the place and released him.

but it was only five months after his arrival that war broke out, and—bad as were the results for the Papal army—it may fairly be claimed for de La Moricière that during those five months enormous strides had been made.

It is not in the province of this small memoir to give more than a very brief sketch of the Papal army, but its most salient characteristics were the following: about ten different nationalities, and at least three official languages, among which English did not figure; everything in the process of formation; nothing ready; very little money; and an almost complete lack of trained men, as is proved by such appointments as that of the poor old Comte de Quatrebarbes, a country gentleman who had done no military service for thirty years and knew no Italian whatever, but was invited to be governor of Ancona, and held that post throughout the siege;[47] and, perhaps worst of all, the unceasing hurry owing to the feeling that time was short.

Throughout all that summer polyglot batches of recruits continued to land in Italy until the army totaled about 18,300,[48] but of these some 6,300 were garrison troops, leaving only about 12,000 available for the field force. The strength of the various arms was as follows: about 16,000 infantry, of whom there were several thousand either untrained or only very partially trained; about 400 cavalry and about 1500 gunners, including garrison artillery; there were only five field batteries

---

47   As he reappears in our narrative, we may add the astonishing fact that he seems to have been rather successful, and to have been perhaps the only foreign officer who achieved some popularity with the revolutionists. He was a gallant old gentleman with absurd views, a very narrow legitimist; he believed in the divine right of kings, in the certainty of the Pope's victory, sooner or later; in the absolute necessity of his officers being of old family. But at the same time he seems to have been fairly practical, extremely open-hearted and cheery, courageous under shell fire, and very kindly, so the people liked him.

48   They have been given as 27,000, but this seems to me to be merely a paper total. De La Moricière only admitted to having between nine and ten thousand available for his field force. The totals given above are arrived at after comparison of many military returns found in the State Archives of Italy.

Sketch of the Papal Army

of six guns each, with only four horses apiece; very few ambulances or wagons. The Pontifical force was probably about as far advanced as a Kitchener's division might have been after four or five months training during the Great War.

The armament was bad; the year 1860 marked a period of change in the weapons of war, when smoothbores were giving way to rifles both in artillery and small-arms, but owing to lack of assistance de La Moricière had not been able to keep up with the times. He had no rifled guns in his army except those which he had made for himself during this short period, and he only succeeded in finding infantry rifles. for two and a half battalions (one Italian and the other the 2nd Austrian Bersaglieri) and three companies of Voltigeurs; the Swiss Carbineer battalion had good Swiss carbines and the Franco-Belgian half-battalion (the "Tirailleurs") had good Minie carbines; but the remaining 10,000 or more of his men had only the old muzzle-loading smooth-bore muskets, which were often extremely bad. One of the Irish veterans has told me that two or three shots. a minute was the most rapid fire obtainable from them, and another "that they would hardly hit anything you aimed at," and one notices that the *Wanderer*, a Viennese newspaper, published an article or two during this summer of 1860, roundly abusing them.

In this curious army there were men of almost every white race; in most of the units there were two languages spoken and in some of them far more. The numbers of the fighting men supplied by each nationality, including not merely the national battalions, but any odd batches all over the army, were, quite roughly as follows:

Italians (approximately) ...	6,500
Austrians (approximately) ...	5,000
Swiss (approximately) ...	3,500
Irish (approximately) ...	1,040

Belgians (approximately) ...	610
French (approximately) ...	530

Besides the above there were Poles, Czechs, and others who were good fighting men but whose numbers are not known.[49]

Of the Italians there is very little to be said: owing to the growth of national sentiment they were almost useless. They consisted of two regiments of the line of two battalions each, and also two battalions of Cacciatori. These were smart enough on parade, and apt to treat the raw blundering foreign recruits with a certain amount of contempt, which was freely returned by the latter, who were quick enough to perceive that these smartly uniformed regulars had not got their heart in the cause; and so it proved when the war broke out. Yet on two or three occasions the Italians fought extremely well; and one must perceive that it was almost impossible for them to have any desire to join this polyglot agglomeration of half-drilled foreigners in shooting down the Piedmontese, most of whom were of Italic race—and still less their own neighbors and friends the Milanese or the Bolognese who had lately joined northern Italy.

The Austrians were perhaps the most serviceable troops in the Papal Army because they consisted to a large extent of regular soldiers, and were commanded by Baron von Vogelsang, a colonel well known in the Austrian service. With them, too, were many officers of good family and training, who had resigned their commission, with the approval of the Emperor, in order to take part in this campaign—free to return to their regiments as soon as it should be over.[50] These Austrians were

---

49   These numbers are obtained not by merely accepting the predominant nationality of each battalion, but by including as far as possible the batches of men dotted all over the army.

  For various reasons recruiting was stopped in France sooner than elsewhere but for that order there would have been more Frenchmen in the Papal army.

50   Baron Brackel, April 21. A.S.R. Com. Gen. tr. pon. 3151.

easily organized in five[51] battalions of the type apparently of Jägers, and soon afterwards they were given the name of Bersaglieri or Marksmen. Of these, the 2nd, under Colonel Fuchmann, was an exceptionally fine battalion. The story is told that its first draft of men consisted of eighty trained soldiers with medals. And—according to Bishop Pelczar—it included a large element of young Poles, who were ardent Catholics and excellent fighting material.[52]

The Swiss were far below their traditional standard; and as the name of the Switzer has always been a synonym for courage, and as, moreover, the Swiss battalions in 1848 had freely sacrificed their lives for Pius IX,[53] it is only fair to them to say that in 1860 they were officially designated as the "Foreign" regiments—not Swiss—and contained men of almost every known race; in several of them there were certainly not more than fifty or sixty percent, of real Swiss; the rest were Germans or Italians, with sometimes a few Belgians, French, and others.[54] This, perhaps, accounts for their having very little esprit-de-corps, which was proved, not only during the war, but by the enormous numbers who deserted for months before it began.

The next on our list are the thousand or more Irish, who, beyond all others, had a distinctive character of their own. Although they had not many trained officers, they insisted on

---

51   The 5th Bersaglieri was never completed: it was still only a half-battalion when war broke out.

52   On the list of officers of this battalion one finds Captain Prosig in whose command the Irish companies were included during the siege. Also the afterwards well-known names of Major Fuchmann and Count Goudenhoven; A.R.S., Com.-Gen. tr. pont. 3154. One great advantage was possessed by the Austrians: their government allowed them to have regular depots in Austria, a privilege denied to the French, Irish, Swiss, Belgians, etc.

53   In the war of 1848 they proved themselves the best troops in the Papal army, although fighting for a country which was not their own, against Austrians, men of their own language. When finally included in the general surrender at Vicenza, they sent their colors to General Durando, asking him to keep them beside him so that the enemy might not insult them or drag mem through the streets, after so many of their comrades had died for them.

54   V. De Mérode's letter on this point. Aug 21—Ib 3151.

being formed into an Irish battalion, which became officially known as the Battalion of St Patrick. A green Zouave uniform was devised for them and it still appears on the plates of Papal uniforms as being that of the Irish, but as a matter of fact very few specimens of it ever saw the light of day. Like everything else connected with the battalion, its uniforms had not yet been served out to it when the war began, and so were never worn, except perhaps by some of the officers who bought theirs privately from a tailor.[55] In such matters as this the Irish were extraordinary unfortunate, owing to their being the last arrivals; similarly their boots, pouches, rifles, haversacks, underclothes, prayerbooks, etc., etc., had not yet been served out before the war began, and they were never served out.

The fact was that a great deal of disillusionment remained in store for the Irish, especially for those among them who had served in the British army; here in Italy, instead of a shilling a day, good food, and fair sleeping accommodation and water supply, they received a penny halfpenny a day, food that did not always suit them, had very few means of keeping clean, and slept among fleas and mosquitos. The men of the Irish Constabulary, who had resigned their good pay and their right to a pension, must have felt that they had made a great sacrifice. One records these things not as a reproach against the unfortunate overworked Papal officials, but as an answer to the constantly repeated accusation that the Papal troops were mercenaries who had come to Italy for the sake of gain. One may feel that Pius IX probably made a regrettable mistake in ever raising these foreign troops at all; but his rank and file

---

[55] Mr. Crean told me that they were not yet served out to the men at Spoleto; so did Edward Dunne, who was there as a private, but he believed that they had arrived at the store. Mr. Mackey and. Mr. Kenny, who were in Ancona, both told me that the distinctive Irish uniforms were never served out there, and described the old Austrian uniforms which they wore. Others knew nothing of the green uniforms. And what the veterans say is fully confirmed by De La Moricière's order on the subject, and the reply from the Sub-Intendente Ferri, A.S.R. M.A., Com. Gen., 3151 Fasc. 1.

must have joined him from honest motives because no one aware of the conditions would have enlisted in the Papal army for gain: in 1860 it was an extraordinarily unattractive service, as was proved by the fact that hundreds of men deserted from it.[56]

The drafts of Irish volunteers continued to arrive in Ancona throughout all that glittering summer in batches varying in strength from about twenty to about a hundred and fifty; their difficulties were far greater than those of any other corps because they alone found nothing ready to receive them. The Swiss, for instance, had a standing military organization in the Papal State; the Austrians contained a high proportion of trained officers and men; the French consisted of two small corps d'élite; and the Italians were in their own country. But the Irish had every single thing to prepare from the very beginning. They had to tell off their companies, appoint officers and N.C.O's, and start the whole organization from the cookshop to the orderly room, from the lance-corporals to the second in command; and this without many trained men and constantly compelled to make bricks without straw; and finally—in the majority of cases—with only about two-and-a-half months' instruction, from the date of their landing.

Thus the Irish volunteers never in reality reached the point of being a fully-equipped battalion. They had four companies in Ancona and four in Spoleto, but some of their recruits had only been with them six weeks or even less, and some of the men had no uniforms even of the French or Austrian type, and none of them had much kit. De La Moricière took one reduced company about one hundred and ten strong, with his field force to Castelfidardo, but even this small number was short of many necessaries, notably of their pouches: they were obliged to carry their cartridges in their haversacks together

---

[56] I am speaking now of the Papal army of 1860. The Papal Zouaves (1860–1870) lived a happy enough life, through dangerous.

with their rations.

Nevertheless, in spite of all difficulties, these companies produced quite a remarkable number of officers who afterwards did well. Their names and careers will be found at the end of this book.

The French element in the Papal army, though numerically small, was undoubtedly the dominating factor in the service, because it included nearly all the officers in command; moreover, it is a remarkable fact that it represents almost the last appearance of the French noblesse on the field of European history—and no one can deny that they made a gallant showing from this date onwards until the day in 1870 when, at Longwy, they were wiped out by the Prussians. During this war of 1860 they poured out their lives in the true historic manner of the Maison du Roi; when the list of killed and wounded was read out after Castelfidardo, one of the Piedmontese generals listening exclaimed: "Why, it sounds like a list of the guests at a court ball of Louis XIV!" Almost every noble family of France had sent a son to the field.

The French contingent was divided into two units: the Volontaires à cheval or Guides; and the Franco-Belgian Tirailleurs who after Castelfidardo developed into the Papal Zouaves.

The Escadron des Guides consisted of forty-one young nobles[57]—a good many of whom had probably French legitimist aims in the back of their minds—under command of Count Gaspard de Bourbon-Chalus, its two lieutenants being Comte Auguste de Gontant-Biron and Vicomte de Saintenac; and the rank and file were like unto them—all noble. Every man paid for his own horse and his own kit, and cost the government nothing. They were the corps d'élite of the Pontifical army and were attached to General de La Moricière himself for staff work.

The second French unit, the Franco-Belgian Tirailleurs,

---

57 Bittard des Fortes states them at 70.

consisted two-thirds of Frenchmen and one third of Belgians, and although it was only a half battalion about four hundred strong, it became undoubtedly the best emit in the Papal service. It was composed very largely of sons of the French aristocracy, but with them there was also an element of heavy, obstinate, Belgian peasants and of sturdy Bretons. To this war, Celtic Brittany supplied a finer contingent than any other province in France, or indeed anywhere. De La Moricière was a Breton, so was Count Becdelièvre, colonel of the Franco-Belgians—and so were six out of the twenty-one dead bodies identified after the battle of Castelfidardo[58] as belonging to the half-battalion. Out of its seventy-one casualties definitely tabulated on that day (of whom nearly one-third would be Belgian) some twenty-one were Bretons, or almost half those of the French contingent. Their feeling is expressed in their songs:

> Ils étaient sur une montagne
> Près de Castelfidardo
> Ca me rapelie aussitôt
> Les côtes de ma Bretagne
> Et je me dis comme ça
> Dieu sait si tu les r'verras...

With them ends our short review of the Pontifical army of 1860, and truth compels one to state that it was a very bad army, mainly because a large proportion of the men composing it had no enthusiasm in the cause and no readiness to offer their lives for it; if attacked by the Piedmontese, they regarded their chances as hopeless. And it was in the old Italian and Swiss element that this feeling was strongest; the young battalions in whom the volunteer spirit prevailed were better than those of the regular Pontifical corps.

---

58   Its losses were about 145 killed and wounded out of a total of 270 engaged in this action, but some of the dead could not be identified.

This war was undoubtedly a struggle between two great principles: those of modern nationality and the historic conception of the Church; and nationality proved itself the stronger. Indeed one may say that it was the guiding spirit of the nineteenth century, the only principle for which almost every man was ready to offer his life.

But even within the Papal army itself one finds that nationality is the truest predominating influence; when the war begins it is the most national units, the French, Irish, and Austrians, which show up best; after them come the so-called Swiss battalions, in reality composed of people of almost every nation, and they surrender easily because they have little esprit-de-corps; finally we have the Italians, in many of whom their national sentiment was actually a negative force; and the really astonishing point about them is that on several occasions they did well.

CHAPTER IV.

The Secret Work of the Revolution and the Press-Campaign.

From the historical standpoint this chapter in the story of our Irish volunteers has an undoubted importance, because it helps to throw light on a vexed question, the true character of the Papal army of 1860. That army has been constantly derided and abused by anti-Papal writers, and systematic efforts have been made to prove that it consisted of ne'er-do-wells, blackguards, and plunderers enlisted to oppress the people; and so often have these things been said that they have to some extent passed into history. This misrepresentation is due to the fact that the Papal question in Italy has remained a living political issue of extraordinary bitterness, even down to our own day, and that the foreign soldiers who offered their lives for Pope Pius IX have been especially easy target; there was no one to defend them because their own sympathizers are not proud of having been compelled to enlist foreigners.

When week after week from April to August, large batches consisting of hundreds of Papal recruits—Austrians, Swiss, Irish, Germans, Belgians, French, Poles, with a few[59] Czechs, Spaniards, and others—were landed at various ports in Italy,

---

[59] I give these nationalities in order of their numerical strength; but the most, numerous contingent was the native Italian. Almost one-third of the Papal army consisted of inhabitants of the Papal state.

they very soon found themselves faced with disillusionment; they had unwittingly stepped into a hornets' nest; on their own side there was hardly any preparation for receiving them, whereas in revolutionary circles everything had been secretly organized to prevent the formation of a Papal army. Below the surface the whole country was covered by a network of revolutionary committees, directed from Florence or from other towns outside the state and ultimately in touch with Cavour himself.[60] Everywhere there were bodies of conspirators communicating with each other by cypher, sending news, spending money, distributing arms, and, in fact, leaving no stone unturned in their effort to thwart the government. The Pope, of course, was not without his supporters; he was beloved in the country districts, and the country districts contained about five-sixths of the population.[61] But the Umbrian peasant in 1860—as even to-day—was practically a negligible quantity, more interested in his vines and small field of maize than in politics.

The work of the revolutionists might be roughly classified under three headings. They worked:—

> (1) By initiating risings assisted by raids from over the border;
>
> (2) By impeding the military work of the government: bribing contractors, organizing and assisting desertion, sending out reports on all points of interest;
>
> (3) Through the Press, of which the chief organ was the *Nazione* in Florence, a clever and important paper only recently founded in order to assist the cause of United

---

60   The works of the Marchese Degli Azzi (anti-Papal), also the great body of research in the Archivio storico Risorgimento Umbro, and *La Liberazione di Perugia*, a historical review.

61   Major Corvetto gives a list of towns and their population, which shows them to contain about one-seventh of the inhabitants, but his list is not quite exhaustive. Corvetto: *La campagna di Guerra nell' Umbria e nelle Marche*, p. 45. This list is quoted by Cellai: *Fasti Militari*, Vol. IV., p. 485.

Italy. The Press worked in conjunction with the committees: in every town there was a secret correspondent who sent in his report about the soldiers quartered there, stating their numbers and armament, and adding any stories or slanders that he could collect or invent against them.[62]

With the first two of the above-named headings our Irish volunteers are not concerned, because they never in reality got beyond the stage of being on the barrack square: some of them did not arrive until the end of July, only six weeks before the war began on September 11th; and, in their case, the usual attempts to promote desertion were apparently abandoned after two of the promoters had been rough-handled. But the third heading, namely the Press-campaign, was for them a very serious matter, doubly serious, because they had to face two press-campaigns, the Italian and the English, the latter of which was backed from time to time by the British Government.

Of course Propaganda—the right to slander the enemy—is a recognized weapon in modern war. All nations use it—ignoble as it appears. So that no one would blame the revolutionary newspapers overmuch for what they wrote about the Papal army in 1860; but there certainly are other people who are very much to blame in the matter, namely certain modern historians who, in order to add excitement to their book or to prove their case against the Papacy, have repeated the absurd accusations made in 1860 without inquiring whether there was any foundation for them or not; and it is the statements of these last-named writers which will be noticed here, because they have grown into a customary defamation of the Papal

---

[62] Copies of many of these reports are in my possession, and also the key to their cypher, which had baffled all efforts to unravel it. These I was allowed to see through the kindness of the Marchese Degli Azzi, whose historical works are well known.

army and thus indirectly of the Papacy itself.[63]

The propaganda campaign was one of extraordinary violence, and even more unscrupulous than anything that one has seen during the last fifty years. This violence was due partly to the exasperation of the nationalists at being checked in the supreme crisis of their history; but it was mainly due to the fact that a war against the Papacy was inevitably a war of beliefs—in spite of all that could be said about not interfering with the spiritual power. Thus we see that Pius IX had excommunicated Victor Emmanuel: it became, therefore, the object of the opposite side to discredit Pius IX in every possible way, and one of the most obvious methods of doing so was by representing him as an oppressor and the Piedmontese as the deliverers of his people. The accusation was so far true that for ten years Pius had been unable to keep order within his state without the assistance of Austrian troops. In fact, to the present writer there seems no kind of doubt that the Papal Government stood condemned as an anachronism,[64] and that—quite apart from the necessity of forming a United Italy—it had long been doing more harm to the cause of religion than good. But that affords no reason for believing all the slanders written during the press-campaign of 1860, and absolutely no excuse for reproducing them.

The easiest target for the journalists was the foreign soldiery of the Pope, and, broadly speaking, their presence was a just cause of complaint. True, the Papal state was an international unit relying for its defense on the powers of Europe and on Catholic sentiment, but in practice this meant that it introduced foreigners in order to fight Italians. Still, as Cardinal Antonelli pointed out, the employment of mercenaries was then a common custom throughout Europe, even for rulers

---

63   It is pleasant to see that most of the latest Italian historians have abandoned this attitude.

64   Editor's Note: O'Clery and other pro-papal historians on the Risorgimento would push back on this point.

who had no international status and no Church to safeguard. And the Piedmontese certainly raised no objections when Garibaldi enlisted Englishmen or Hungarians for his invasion of Naples; perhaps because nearly all their own generals, Fanti, Cialdini, and all the Garibaldian leaders, had spent their younger years fighting in foreign armies. Moreover, Italy was not yet a nation, and at the beginning of 1860 it still seemed doubtful whether she would achieve her unification.

It would have been hard for Pius IX, bound as he was by his coronation oath, to let the Papal state go without making a final appeal to Catholic Europe, and at the time when he first made this appeal, early in 1860, there was no actual certainty that the desire for a united Italy would so soon become an overwhelming force. But all through the summer of 1860 Italian national sentiment made enormous strides, mainly, I think, owing to Garibaldi's wonderful exploits in Sicily; he supplied the people with a fighting tradition of which any nation might be proud.

To return, however, to our immediate subject, the press-campaign against the Papal army, one or two specimens from the *Nazione* will suffice to show its value from the historical point of view.

April 12, 1860, speaking of the Swiss battalions in the Papal service, it says:

> The imprisonings of the people were accompanied by some official searches which proved fruitless, as usual. Then the column [of soldiers] moved on to Gubbio; and it was a fine sight to see, with its mounted Gendarmes riding in front and in rear, so as to prevent desertions; there were numbers of patrols, too, guarding the main road that leads to the frontier. On the following morning 800 more Swiss arrived from Pesaro

> on their way to Perugia. This miserable collection of ruffians kept sprinkling tricolor cockades along the way in order to excite our population to acts of imprudence, but the people saw through the trap and took the greatest care to keep out of it. Then these defenders of religion and order began to ill-treat the peaceful citizens; and woe betide anyone who dared even to breathe; in one inn, for instance, the soldiers drank a great deal of wine without paying for it, broke the measures, and, when the proprietor complained, they bound him ana took him off to Pesaro with another unfortunate inhabitant.

This description of the Swiss regular infantry speaks for itself; the writer is partly redeemed by the fact that he has an incipient sense of humor; but he expected of course to be taken seriously.

July 20, speaking of the Swiss and French, the *Nazione* tells us:

> The first to pass by was the Swiss battalion which we call the battalion of the assassins of Perugia; it was commanded by Colonel Pimodan. Some of these people were heard to say that within a few days they would be at Bologna where a feast-day had been arranged for them similar to that at Perugia; others complained bitterly of the discomforts of their life and concluded that they were being sent to the slaughter... others added that when they got a safe chance they would show everyone which side they meant to turn their arms against, unless they could make their escape to the frontiers first. I leave all comments to my readers...

The other corps that passed through here was a horde of thirty-five mounted men which calls itself the Guides—a miserable collection of French legitimists and republicans captained by a bastard of the house of Bourbon. I say nothing of the ridiculous manner in which they entered the district, nor of their mounts, but need only tell you that they did not even know enough to get through correctly the usual elementary "dismount"; that is of no importance. But these new paladins behaved almostly exactly like brigands: when they had sat down to mess in a lodging-house improvised for them by the Gonfaloniere, they ate and drank as much as they wanted and then called for the bill; but then, without even looking at it, each of them threw into his plate a twenty-bajocchi piece [value about one franc] and departed regardless of the complaints of the innkeeper who was about four scudi out of pocket, besides the use of the plate, wood, oil, etc.; and some of them were not even ashamed to carry off some fiaschi of wine. After luncheon came the turn of the poor communal secretary, a man named Domenico Buoncristiani, more half-witted than offensive. This unfortunate fellow—merely because he had not been quick enough in attending to their wants—was seized by the necktie, a pistol was pointed at his chest with threats that he would be killed, and finally he was obliged to shoulder the straw for the horses' litter... A trustworthy person from Viterbo assures me that they refused to pay Schienardi the innkeeper there a sum of fifty scudi due for their messing, and that they abstracted four silver services.

A story like this is impossible to believe, though it may, of course, have had some slight foundation of truth. The French Guides were a corps d'élite, especially attached to General de La Morcière as a kind of staff; they were all nobles or sons of nobles; they served without pay and provided their own horse and uniform were expected "to be in a position to spend 4,000 to 5,000 francs a year of their own."[65] It seems hardly likely, therefore, that living under the eye of the General they should have publicly disgraced his staff; and still less likely that they abstracted the silver forks and spoons.

The stories against the Austrians are even worse than the above: on June 7 they are accused of striking men and women at Perugia; on June 14, of robbing peasants and carrying off girls into their barracks by force at Spoleto; and on June 24, of robbing shops and attacking women in Viterbo; and there are many other allegations, which seem entirely incredible when one remembers that the Austrians were nearly all regular soldiers with regular officers of good education and position, serving under de La Morcière, a Lieutenant-General in the French army. For his own sake he could not possibly have tolerated excesses such as these.

All this is propaganda—and propaganda of the crudest kind. It would not be worth mentioning if it had not to some extent passed into history.[66]

On August 17 we get a broad statement of the *Nazione's* views about the Papal question:

> What is the aim of the Roman Curia? To flagellate its subjects; to oppress them with a govern-

---

65 A.S.R. Com. Gen. tr. pont. 3151.
66 Anti-Papal writers quote such stories to prove their case. Modern Italian writers, however, do so less and less; but in the year 1910, the 50th anniversary of the Italian nation, all Italy was deluged with writings full of these and similar anecdotes to celebrate their delivery from Papal rule.

ment that tears their sons from the arms of their parents; with a government that rewards and glorifies the sacking, the assassinations, the massacres, the rapes of Perugia, and if it were not ashamed to do so would be prepared to beatify the heroes of such infamies as these.

From the beginning of April until the middle of September, when war broke out, the *Nazione*—and plenty of other papers as well[67]—gives us an uninterrupted stream of accusations against the Papal army and the Papal entourage. It was in the month of July that they came thickest, and in that month there are only about seven or eight days on which a fresh accusation is not printed against one or other of the Papal battalions. The following are a few specimens:

July 1—An article complaining that Italian officers in Papal service are being "sent to jubilation" to make room for foreigners. That the Irish were drunk and stole things; that the Italian Papal Gendarmes had brutally charged a crowd without provocation at Frosinone.

July 2—That all the Italian Papal troops except the Gendarmes had been won over to the revolution. That the German and French-speaking regiments were at loggerheads with each other and with the Italians.

July 3.—An article against the Irish for disorderly conduct in Macerata.[68]

July 4.—An article against the Irish for rioting in Rome.[69]

July 7.—An article against the Pope and his police.

---

[67] It would be very easy to quote similar articles from many other papers, as for instance, the *Monitore Toscano,* the *Perseveranza,* etc. But, I understand, the *Nazione* was the central recipient of the reports sent in by the revolutionary committees all over the country.
[68] These are the two most serious accusations that were brought against the Irish volunteers, because there was an element of truth in them. In neither case were there any serious results.
[69] Cavour's letter of August 31st to Persano. V. Persano's Diary.

July 8.—An article relating how three Perugians had been arrested by the Swiss; some complaints from Gubbio about the Papal police.

July 9.—A furious article about the beating and ill-treatment of the people in Camerino by the Italian Papal police.

And so forth. The *Nazione* brought fresh accusations on almost every consecutive day throughout the next two and a half months, until war broke out on September 11. In only just a very few instances is there any foundation for what it says. The object, of course, is perfectly plain: it is merely paving the way for intervention. Cavour had already decided on intervention.9 Four days before the outbreak of war—that is to say, on September 7th, the day when the ultimatum was actually written—we find a furiously indignant article in the Nazione describing the oppression of the Papal provinces, and adding "the daily martyrdom which they are suffering must be brought to an end."

The aim was to represent the Papal state as suffering martyrdom and the Piedmontese as being deliverers. Cavour's ultimatum, which reached Cardinal Antonelli on September 10th, made this claim quite openly: "The conscience of King Victor Emmanuel does not permit him to remain an impassive spectator of the sanguinary repression whereby the foreign mercenaries are suffocating in Italian blood every manifestation of Italian sentiment."

And on the following morning General Cialdini crossed the frontier issuing a war proclamation which began:

"Soldiers!—I am leading you against a band of drunken foreigners whom thirst for gold and a desire for plunder has brought into our country."

This proclamation might be called the culminating point of the press-campaign.

CHAPTER V.

THE DIFFICULTIES OF THE SOUTHERN COMPANIES.

OUR next enquiry must be: How did the Irish fare when they landed in successive batches in this hornet's nest?

In reality, one must divide them into two halves which never had even a glimpse of each other throughout the campaign:

(1) Those who were sent southward; of these, some were first at Macerata, and at Rome, but they were all eventually assembled and formed into four companies at Spoleto.

(2) Those who were left in the northern seaport of Ancona, and were formed into four companies there; this half-battalion, once it had settled down, had a very satisfactory period of training.

Unfortunately, the same could not at first be said of all the scattered contingents that belonged to the southern detachment.

It is the difficulties and the irregularities of one or two of those southern batches which earned rather a bad name for the Battalion of St. Patrick in some quarters—even in the sympathetic pages of Mr. George Trevelyan. For that reason it is necessary to deal with the events in question, although in reality they have no great importance.

Broadly speaking, the volunteers suffered from three

different classes of difficulties: firstly, from the opposition of the revolutionists who were determined to prevent the formation of a Papal army, or, at all events, to discredit the Pope for forming it; secondly, from bad organization by the authorities; and thirdly, from the opposition of the British newspapers and the disapproval of the Government.

Under the first heading, one may admit at once that our boys were absolutely "a gift" to the revolutionists. One can picture the successive batches of newcomers, arriving in the country districts, many of them raw recruits, still wearing the clothes in which they had lately worked on the Kerry mountains, or strolled the streets of Dublin, or Cork, according to their station in life, now swinging along the roads to Macerata, with songs and laughter, splendid physically, but as yet unsophisticated and unable to understand a word of the language around them; marching, therefore, blindly amid the network of secret revolutionary committees who were bent (and as we now realize, naturally bent) on discrediting them by every means in their power: their smallest frolic was represented as a riot, their every action, even their untrained appearance became the subject of jibes and insults in the revolutionary papers. Yet, at the same time, in some places and among certain classes they were received as friends, while in others the attitude of the people varied from one extreme to the other according to the political barometer. In fact the whole situation was puzzling; and even now, fifty years afterwards, it is quite a common thing when talking to these old men, especially with the less educated, to notice still a naive expression of surprise at this strange reception by a population whom they had come to defend: they had expected to find everyone devoted to the Pope and to be received as deliverers.

On their first arrival the volunteers tried to show that they were friends, and, not knowing a word of the language, they would herald their presence in a new place by giving a great cheer for the Pope; but this effort usually proved a failure. At other times they would go into a trattoria to have a glass of wine, whereupon some of the guests would perhaps get up and go away: in fact they soon began to realize that quite a tangible section of the population regarded them as enemies. As this fact dawned upon them it undoubtedly produced some reaction on their own sentiments: there were some of them who began to wonder if it was worth while dying for a people who apparently did not want to be died for; but the great majority had made up their minds to defend the Pope, and the only effect of these incivilities was to make them rather more independent, more careless of those around them, but equally determined to carry through their adventure to the end.

A certain amount of irritation there undoubtedly was, but—and this is an all-important distinction—it never found vent in any acts of revenge against the people. It is a very remarkable fact that though the Irish volunteers are bitterly, furiously attacked in the Cavourian Press, they are never accused of ill-treating any man and still less any woman in Italy. Accusations of this nature were constantly made, and sometimes with truth, against the Swiss, Austrians, and French, and against the Italian Gendarmes, but never against the Irish. Against them the principal points singled out by the propaganda Press were that they were unkempt and undisciplined and that they stole articles of food when on the march:[70] and

---

[70]   As regards their being unkempt, this is only true of a few unfortunate individuals. In the ranks there were some poor men, peasants who had left Ireland in the clothes in which they worked in the fields, expecting, naturally, to be served out with uniforms as soon as they arrived; but weeks went by without their being given the uniforms. And one of the principal grievances of the volunteers was that when they first arrived there were so very few facilities for keeping clean.

The following description of the type of young men who volunteered was given in the *Freeman's Journal,* of September 14th, 1860, at a time when it was matter of common knowledge: "It is notorious that they are for the most part lads who belong to middle and upper sections of the mercantile class; educated, intelligent, unused to hardships, they fearlessly rushed to the rescue of the Head of their Church."

V. Also the report of the Austrian authorities on the detachment which included

it is undoubtedly true that they were undisciplined at first, but as for stealing food, this is an accusation pointed at all the Papal corps in turn, and seems to have been due to the absence of a proper requisitioning service.

In such conditions as these, there was hardly a Papal battalion that escaped without having some periods of serious disturbance during the summer, and among the Irish the trouble first broke out at Macerata, where rows were nearly always in progress. Macerata was the bitterest town in the Papal State, except perhaps Perugia; already the Austrians had been at loggerheads with the populace there, and the Papal Delegato, Monsignor Apponyi, perhaps true to his good Hungarian name, had sided against them as he now did against the Irish and afterwards against both Austrians and Swiss.[71]

In this town, there happened to be a detachment of several hundreds of Irish recruits of the rawest type. This batch most undoubtedly got out of hand, as was not very surprising: it included most of the peasant element among the volunteers, with little or no steadying leaven of old soldiers; the men were civilians, not yet enrolled, and there were no real officers, for though two or three had been named their commissions were not yet signed; so that obedience was voluntary, and the N.C.O's had been hastily selected out of the crowd. But in spite of these disadvantages, it is worth noting that this same batch of recruits, during their stay in Austria, had received an excellent report from the authorities there.[72]

At Macerata, however, some of our volunteers—and, rightly or wrongly, the batch known as the Kerry boys got the credit of it—who had never seen wine before, on finding

---

most of the peasants.—A.S.R Com. Gen. tr. pont. 3151.

[71] V. Guerra's telegram of Aug. 2nd. and De Courten's and Apollonyi's telegrams Sept. 3rd Also dde Ka Moricière's of Sept. 4—A.S.R. Com. Gen. 3155, and 3154.

[72] V. Count Bracket's report—A.S.R. Com. Gen. tr. pont. 3151.

The Difficulties of the Southern Companies

themselves suddenly thrown into a country where it is cheap and abundant, proceeded to sample it rather recklessly, just as did some of our British soldiers in the year 1918. The results were unfortunate; on two occasions they resisted the guard that was sent to bring them back to barracks; and when one popular volunteer was arrested they refused to leave him to his fate or to go home until he was released; some stones were thrown, some windows were broken, and the townspeople naturally and rightly complained.[73] It was an episode unsuitable to defenders of the Church and, doubtless, a bore to the citizens, and it frightened the police. But still, one remembers that similar events took place in every town in England during the Great War—and in some towns they took place almost every Saturday night—without our being greatly troubled about them.

But the authorities there, the Papal Delegato and the Town Major, were considerably scared and entirely unequal to the occasion; it was due to their alarm that de La Moricière was obliged to follow the one course which ought to have been avoided:[74] they actually got the Macerata contingent started off to march right across Italy to Rome, apparently caring little for results, provided that nothing fresh occurred at Macerata.

---

[73] V. Telegrams between June 18th and July 3rd—Ib. 3154; Fasc. 39.

[74] It does not seem that there was even one man seriously hurt in this row, yet the Delegate applied for a company of regular soldiers to assist the police in keeping order. He had already done so when there were only 140 recruits at Macerata. De La Moricière evidently thought this unnecessary for, in answer; he wrote the following telegram to Colonel Guerra, C.O., Ancona: "Do not send the company without my order; pay no attention to the Delegato; in any case, I shall wait until Keller applies to me for it." (Telegram, June 24th, . . . A.S.R., Com. Gen., tr. pont, 3155. Fascicolo 44.) The volunteers themselves though they fully admitted that later on at Spoleto there was a period of serious discontent, regarded this Macerata episode purely as a joke. They attributed the disturbance mainly to a "Kerry boy" who had started beating a drum at midnight and shouting "Garibaldi," and had thereby caused a garrison alarm, and they added that this man, Murphy, was afterwards the hero of a splendid act of courage during the fighting at Perugia. Among this batch at Macerata there were certainly some of the discontented section, but their grievances had not yet come to a head.

It was this march from Macerata which was responsible for most of the accusations against the battalion, although in reality it concerned only a particular batch of the southern half-battalion.

They were sent off from Macerata in at least four successive batches; the first, 154 strong, started on June 16, reached Nami on the 21st, and went on by river to Rome; the second batch, 133 men and four officers, started on June 18, arrived at Narni on the 24th, and reached Rome on the 26th; it seems to have been the least disciplined of the party. Meanwhile, on June 24, the third and largest batch, consisting of 262 men under 2nd Lieut. D'Arcy, started for Rome via Foligno, but was halted at Spoleto which had lately been selected to be the future depot of the battalion.[75] On June 30 the fourth and smallest batch of 63 men marched off under 2nd Lieut. Stafford. With so many detachments of civilian volunteers sent careering all over the country, mainly by night, it was certain that some contretemps would occur, and many absurd misunderstandings; but nevertheless, the first, third, and fourth batches seem to have got through their task wonderfully well. It was the second batch which became rather too reckless.

One must remember that the political atmosphere was exactly like that of a general election—only far more bitter—so that every conceivable item was seized upon for propaganda; and not infrequently items were fabricated. In reality there was nothing genuinely serious to complain of. Many stones have I heard of the march from Macerata to Rome from the lips of the old men themselves, or from those of their sons and daughters, and they give a ludicrous impression of an utterly incongruous situation: the 133 Irish boys marching along in

---

[75] This large batch was under command of Second Lieut. D'Arcy, who had never done any soldiering, and, according to the returns, was not yet 17 years of age. Nevertheless he got through the march wonderfully well. It was not until this batch arrived at Spoleto that their difficulties began. There are a great many telegrams about them during these days, but their movements are hard to disentangle.

the very highest spirits, at times completely out of hand—as was to be expected with only three officers, all provisionally named, and a few haphazard N.C.O's—amid a population which was considerably afraid of them, and a network of secret committees which reported with profound solemnity their most trivial actions as breaches of military discipline, and sent these reports to be published in the great *Nazione* newspaper in Florence.

In the villages there were occasional disagreements: but it is only fair to the volunteers to say that they never intentionally purposed any harm or loss to anyone; their adventures were more analogous to those of Lever's hero, Charles O'Malley, in Spain, and just occasionally got so far as to resemble that of Mickey Free with Mr. Meekins, the reporter of the *Bristol Telegraph*. But during this long march by night as well as by day there were many genuine misunderstandings. The following is an instance:—

As the volunteers entered one of the villages on their line of march[76] they saw themselves greeted by faces at almost every door and window. By way of a friendly reply they at once sent up a cheer of welcome; and immediately every head hastily disappeared, doors and windows were closed, and some of the inhabitants fled from their homes; they had been taught to regard the Papal troops as plunderers and mistook the cheer for a war-cry. But here was a difficulty: the volunteers were tired out and required food; the shops were deserted. Finally, therefore, they were obliged to help themselves and leave the money on the counter. But it is evident that if any one single dishonest man, either Irish or Italian, had succeeded in appropriating any of that money there would have been ground for a genuine accusation of plunder fit for publication in every

---

76   I originally read this story in a newspaper; but Mr. Bergin, who served with the Papal battalion tells me that it is perfectly true, but he thinks it occurred on the march to Macerata, and not from that place.

revolutionary newspaper.

The volunteers suffered severely from the heat of the Italian sun at the end of June and beginning of July during their march of about one hundred miles in six days;[77] between sore feet and sleepless nights—for there was very little water supply and hardly any beds—quite a number of them were obliged to go into hospital when they reached Rome. During their seven days tramp their best friend was the chaplain Father McLoughlin, a stoutly-built red-faced padre of about forty-five with an unfailing sense of kindly humor. "He was an Irish Franciscan who, by his general aspect, his high stature and his sonorous eloquence reminded me of the great O'Connell, as I had heard him described."[78] Already at Macerata he had won an immense influence with the men: when he walked down a room they fell in on either side! ready to accept his reprimands if necessary. And throughout the whole of this march he seems to have proved himself quite the most useful among the few officers there.

When the second batch above-named arrived at Rome, it was—as might be expected—a good deal out of hand, and soon signalized its presence by giving rise to an incident which was small in itself but has assumed a certain importance because it reached the ears of the British Consul and was described in his report, and has therefore entered into history by a legitimate door.

What happened was simply as follows: a day or two after their arrival at the end of June, when about three hundred of the volunteers were still in Rome, the adjutant-elect, Mr. Howley, until lately an officer in the 10th Hussars,[79] met one

---

[77] Most of the detachments marched to Narni in six days; from there they were conveyed, I think, by boat. But this one under D'Arcy was stopped short at Spoleto.

[78] Russell of Killough.

[79] He was son of Mr. Howley of Belleek Castle (Co. Sligo), and is said to have left the 10th Hussars in order to serve the Pope. He became adjutant of the battalion.

of his men named Laffan, an ex-medical student, walking about the streets in mufti, so he stopped and questioned him; and Laffan, who was apparently discontented, replied quite openly that the promises made to him had been broken and that he, for one, intended to go home; whereupon Howley ordered another volunteer named Wiseman to arrest him; but Wiseman—and this gives an idea of how entirely some of the volunteers failed as yet to realize military conditions—excused himself on the grounds that Laffan was a personal friend of his own. Mr. Howley therefore ordered two passing Italian soldiers to carry out the arrest, while he himself went off to call the guard; and he returned shortly afterwards accompanied by the guard but only to find that Laffan had defeated the two soldiers and was gone. Wiseman, however, surrendered, and the incident ought to have closed without producing any serious results—except for Laffan. But by that time the news had spread to the barracks that Laffan was being bullied by Italians. Cut off as they were by language from all other units, the volunteers felt it all-essential to stand by one another, and, regardless of orders from the French officer in command, they turned out at once to the rescue. Immediately the rumor spread through Rome that the detachment of Irish volunteers was in mutiny, and the Franco-Belgians were ordered out to keep them in barracks. It might have been a serious position, but the men of the two battalions were on very good terms, and, by a stroke of extraordinary good fortune, at the most critical moment Major O'Reilly suddenly appeared on the scene. He was still in mufti, having just arrived in Rome.

Stepping into the place of the French officer in command, he told them how greatly he regretted to find at the very first moment of his arrival that his fellow-countrymen were, by their lack of discipline, endangering the name not only of their battalion but also of their country.

As he spoke most of the resentment vanished. But it took

some time to restore order. The volunteers, who for some days had been chafing under a foreign commanding officer whom they could not understand, loudly cheered their own major. Order was re-established, and the incident was at an end.

It is the above episode which has told most against the Battalion of St. Patrick, and has told rather unfairly; for, after all, it was over in two or three hours and little or no damage was done. It did not trouble the Pontifical Government much—they had had on their hands a genuine mutiny of the Italian Cacciatori only a month or two earlier—but it took place in the capital city and gave a handle to the hostile press. This and other irregularities were due to the fact that for Irish recruits there was no existing organization of any sort whatsoever: no battalion, no companies, no cadres of officers or N.C.O's; the volunteers were simply shoveled into Italy by the hundred without any decision having been taken as to the formation of their unit. In this respect, they were at a great disadvantage, as compared with the recruits of other nationalities;[80] the Swiss and Austrians were in an entirely different position because they came to a country where there already existed whole battalions of their fellow-countrymen and an organization in being; but the Irish had everything to begin from the very start. It was only on June 29th, that is to say, after the episodes at Macerata and Rome, that the cadres of the Irish battalion were first published, and even then only four companies were provided for out of the eight, and only six officers were named, of whom three received merely provisional commissions.[81] Of the privates, none were yet formally enrolled. Some, apparently, had taken an oath of allegiance before leaving Ireland, but the conditions depicted and the promises made were so entirely misleading that these first

---

[80]   V. De La Moricière's letter, dated between July 13th and 17th, kindly presented to me by the Marquis MacSwiney of Mashanaglass. It was part of the collection of his friend the late Baron Kanzler.

[81]   Order of the day June 29—Com. Gen. tr. pont. 3155.

oaths of allegiance were never held to be binding; in fact the men remained legally civilians, not subject to military orders.

Thus, at the end of June, they were practically a flock without a shepherd. Several hundreds were at Ancona, a few at Macerata, several hundreds at Rome, and D'Arcy's batch was at Spoleto, which was destined to be the depot of the future battalion.

At this point we come to the real difficulty which had to be settled before the battalion could be definitely started on its training, and that difficulty came from within; of the thirteen hundred or more Irishmen shoveled into Italy, the immense majority had come with the most honest motives, believing that they were the defenders of the Church and of religion. We have already noted the sacrifices made by men such as those of the R.I.C., and it was said and believed at the time, that the volunteers of the more well-to-do classes would find practically every profession closed to them on their return home. But together with these genuine enthusiasts, there was a small percentage of men who had come hoping for personal advantage, and they were certainly not long in discovering their mistake. Their numbers, not very large at first, were gradually swelled by the adhesion of malcontents or disappointed men, and this discontented element became, during the first three or four weeks, so serious a factor in the situation that the battalion was never able to make any real progress until the malcontents had been sent home; after which, it advanced with astonishing rapidity.

The problem thus created was one which in reality left the authorities rather helpless. The truth was that there was a great deal to be said on the malcontents' side of the question: they had been enlisted on the voluntary system, which, in the middle years of the nineteenth century, meant that any imaginable lie might be told—and in this instance there is not the slightest doubt that such lies certainly had been told. at all events by the subordinate agents probably accustomed to the ordinary methods of recruiting. Some of the men had come out relying on deliberate promises—such as that of a commission—which could not possibly be fulfilled, and de La Moricière himself admitted this in a telegram of July 19, to the Minister of Arms, in which he said:[82]

> I consider it certain that all this last detachment of Irish has been engaged by means of unreasonable promises, and that they were delayed in Vienna so that they might be given military ranks.[83]

But even the most honestly disposed recruiters had drawn a picture of military life much as they knew it in the British army, whereas life in the Papal army was a very different matter: it was, as yet, a daily struggle to train and organize raw recruits, without time and with very little money in hand; the pay was about a penny halfpenny per day, instead of a shilling; instead of the "pound of dry rotee"[84] of the British army they only received fourteen ounces of bread, and instead of the pound of good meat the ration consisted largely of macaroni to which many of the Irish never really became accustomed:[85]

---

[82] V. also De Mérode; telegram July 14—Ib. 3151.
[83] A.S.R. M.A., Com.-gen. Busta, 3155, Fascicolo 45. He was speaking at the moment of the Irish in Ancona, not Spoleto, but this does not affect the question of promises made before they left home. In another telegram to de Mérode, of July 14th, he added: "The officers including the Major (Fitzgerald) received a higher rank than that which they held in the Austrian Army. I contested their right to these ranks, and they proposed to go to Rome to discuss this question with you." The Papal authorities did not admit the legality of *any* commissions granted before volunteers arrived in Italy. This naturally gave rise to great disappointment among those who had been promised commissions in Ireland or in Austria and had led to the men right across Europe. (V. de La Moricière's telegram of July 11th.)
[84] De Mérode; telegram July 4—Ib. 3151
[85] One of them has told me how some of the volunteers were discontented because, with only a penny-halfpenny a day they could not get enough bread; and others were miserable because, if they bought bread, they could not afford tobacco.

The Difficulties of the Southern Companies 63

yet they were lucky to get even this, for, at one station, the Italian gendarmes were receiving only one meal a day, and said that their straw had not been changed for ten months;[86] and in several others the bread was immediately condemned as dangerous by de La Moricière and the doctors.[87] At first the men were herded together and slept on damp straw which was changed only once in four months and soon became alive with mosquitos and the other insects of a hot Italian summer. In some stations water was hard to get; in others, the contractors were undoubtedly becoming rich at the expense of the soldiers, especially those who were secret sympathizers with the Revolution. We know all these things, not only from the complaints of the men, but from the reports of de La Moricière himself, who tried hard to put an end to them.

It was a great pity that on their first appearance at Spoleto, their future depot, the volunteers should have found all these evils particularly rampant. D'Arcy's batch were the first to arrive, and found the conditions of life practically impossible; so much so, that he sent to the Minister of Arms a telegram which almost amounted to a resignation; and there is no doubt that he was right.[88] When de La Moricière appeared on the scene a week later he went into the question, and on that same evening wrote a private telegram to the Minister of Arms in which he said: "The men are the subject of an ignoble speculation."

Matters came to a head when the whole southern half-battalion was finally collected for training at Spoleto on July 24; about 210 of the men there refused to enlist because that involved taking the oath of allegiance for four years, and they did not feel equal to sacrificing the best four years of their life to service in such conditions.[89] However, of this total only 163

---

86 De La M.'s "Rassegna" April—Ib. 3154 His telegram July 3rd. Ib.
87 A.S.R. Com. Gen. tr. pont. 3154 Fasc. 34, and 40.
88 A.S.R. M. A. Aff. Ris. 1964.
89 Pimodan's telegrams July 24th, 25th—A.S.R. Com. Gen. tr. pont. 3154

actually went home, and m the meanwhile 48 fresh recruits arrived. It is said too that the old soldiers who had served in the British army made very invidious comparisons, and were clear-sighted enough to see that the Papal army was most unlikely ever to become an efficient fighting machine.

But the outcome of this difficulty must be left to a future chapter.

During these troubles at Spoleto a rather significant incident occurred, showing the attitude of the British Government towards the volunteers. A certain private named Denis O'Keeffe while undergoing punishment in cells at Spoleto succeeded in getting a letter through to Mr. Newton, the British Consul in Rome, claiming his protection. On August 7 Mr. Newton forwarded it to the Foreign Office in London, together with a dispatch[90] in which he talked of disorders in Spoleto, adding "that more than one Irishman was cut down by Carabinieri, one, it is said, by Colonel Pimodan himself"; that "General de La Moricière was in favor of distributing the Irish among various foreign regiments; that he is not favorable to Irish officers"; and he (the British Consul) talks of having signed fifty passports "since my arrival here." It is hard not to believe that the phrase about the Irishmen being apportioned out to foreign regiments and the Irish officers treated unfavorably was not designed to discourage recruiting; if this was so, it came too late, because by the end of July recruiting had ceased.

But what is astounding about the episode is that three weeks later O'Keeffe's letter appeared in the *Times* together with extracts from Mr. Newton's report. From thence it was copied into the Italian newspapers, and has lately appeared as historical evidence in one or two of the modem Italian histories

---

Fasc. 40.
90   F.O. Rome. Vol. 81. Consular Report, August 7th.

The Difficulties of the Southern Companies

of the time.[91] This is undoubtedly an instance of the campaign which was in progress to discredit this small national effort of 1860: the battalion was condemned on the evidence of a man whom it had imprisoned.

Mr. Newton denied that he had any hand in publishing these documents, and certainly as a general rule he writes with moderation; but he was in touch not only with the Irish who wanted to get home, but also with one of the two or three Englishmen serving in the companies at Spoleto, and this man sent him stories against the volunteers; and nowadays, sixty years later, these stories have received some credence simply because they came through the British Consul and were reproduced in the *Times!*

There would have been no need to recall the O'Keeffe episode at all had it not been solemnly quoted in one or two of the modem Italian works on the period.

Another curious instance of their difficulties seems just worth recalling. In Spoleto was a very active revolutionary reporter who wrote his secret dispatches under the name of Simone. Several of them I have seen. They were extraordinarily bitter attacks upon the Papal troops at Spoleto, amongst them, of course, the Irish. At the time they had very little importance, although they were published in the *Nazione,* but one of these

---

[91] It is quoted in Degli Azzi's book, *La Liberazione di Perugia;* but since writing it the Marquis Degli Azzi has somewhat modified his opinion about the Papal soldiers; and has stated that fact in the most broadminded manner. *V.* The *Risorgimento Italiano* for 1913, a note under the article, *Gli Irlandesi al Seruizio del Papa,* p. 1.

All the above statements sent to Mr. Newton bear the marks of a hostile informant. They are exaggerations of small episodes. It was true that two of the Irish who refused to enlist, also refused to obey an order, and were struck by the Carabineri with the butt-ends of their rifles; but such episodes occur in every battalion. It was true that originally, the Irish had been anxious lest they should not be allowed to form a national battalion. But it was entirely untrue that such ideas were supported by de La Moricière. He, on the contrary, worked the battalion on national lines and was its best friend.

Mr. Newton appeared astonished when his stories appeared in the *Times.*

has been reproduced in a modern work on the period, and is worth translating as a specimen: The Irish, he says, have an extraordinary form of military punishment: "There is a most singular method of punishment among these people. They wrap a man round in a blanket and two or more others who hold the ends of it throw him furiously into the air a certain number of times, according to the gravity of his crime, and bump him with all force and fury upon the ground."

He had evidently seen some of the boys amusing themselves by tossing each other in a blanket, and had mistaken this game for a military punishment!

Was there ever a battalion formed under more unusual difficulties?

But, at the end of all these dreary details, there is one redeeming feature in the picture:—Although surrounded on all sides by enemies, both secret and open, nevertheless, the volunteers had unconsciously acquired one valuable friend, and one who never failed them, namely, their General. Whatever their faults, these boys were fine material for war, and just the kind of rather lawless material that had appealed to de La Moricière ever since he founded the Zouaves and disciplined the crowd of reckless French recruits, shipped out to him in Africa to be got rid of, and turned them into the splendid colonial battalions of France. As the Irish lads marched along at haphazard through Umbria-they were not very military in appearance, but nevertheless physically they were a fine sight to see—many of them over six foot high, for they included quite a number of the Irish Constabulary and Metropolitan Police, and were largely drawn from Tipperary, Kerry, and other districts, where the men grow large. Clean-limbed, loosely-built boys, with a swinging stride and good blue eyes, a ready laugh, a constant capacity for horse-play, and, above all things, a genuine enthusiasm for their cause—these points were evidently noted with approval by de La

The Difficulties of the Southern Companies 67

Moricière, when, on July 3rd, 1860, he held his first inspection of them at Spoleto. That same evening he wrote a confidential report to Cardinal de Mérode, the Minister of War, and, in the Royal Archives in Rome[92] there exists a copy of this telegram which runs as follows:—

> The G.O.C. to the Minister of Arms:
>
> Spoleto, July 3rd, 1860.
>
> Have seen the Rocca (citadel). Letter following with regard to evacuating it without delay. The men are the subject of an ignoble speculation. Concentrate here all the locked vehicles[93] that we have. Pimodan will come and organize the movement. Civita Casteliana can receive 100 men.
>
> Have seen the Irish. Magnificent troops. Mr. D'Arcy who is in command is very nice, but has never served as yet: among them many old N.C.O's who will make the troops. Five or six men to send home. The Commandant will attend to that. At Spoleto, the population is very satisfactory; also the Delegate.
>
> La Moricière, G.O.C.[94]

---

92   Archivio di Stato. Busta 3154. Fascicolo 40: Telegrammi Luglio, 1860.
93   The Rocca or Castle, at that moment contained some con victs who were to be cleared out of it, before it could be handed over to the Irish.
94   A.S.R. Com. Gen. tr. pont. 3154 Fasc. 40.

CHAPTER VI.

The Half-Battalion in Ancona.

With the story of these four companies, we can deal shortly, because they remained quietly in Ancona, and, after a few preliminary difficulties, spent their time in drilling industriously among the other Papal troops during the nine or ten weeks before the war.

They arrived on the night of July 5, and de La Moricière, being himself in Ancona, at once went down to see them. This batch consisted of Major Fitzgerald and four captains and 538 men.

Here, too, there arose the same difficulty as at Spoleto, for in this case also many unrealizable promises had been made. The terms offered came as a great disappointment to some of the poorer volunteers who had formed plans for sending home bounties, etc.,[95] and had been entirely misled.[96] Nevertheless,

---

[95] *V.* Guttemberg's letter. August 27th, saying that most men wanted to send their whole bounty home at once—A.S.R. M. A. Aff. Ris. 1964.
[96] There is no doubt about this. Recruiting of course was extremely unscrupulous in 1860, and some of the agents in Ireland had undoubtedly exercised the customary license to its utmost limit. The following extract from a letter of de La Moricière to Monsignor de Mérode, Minister of Arms, gives a true account of the case:

"The officers [of the Irish in Ancona] who had served before are generally very satisfactory, especially Lieutenant Guttemberg, who was serving in Bavaria; he was summoned to Ireland by the Archbishop of Dublin, and brought the Irish out here from their own country.

"He declares that in spite of his protests, the wildest promises were made to

493 Irish, out of the 538 at Ancona, signed their engagement form, some others were absent, and only eleven refused to do so and had to be sent home.

De La Morcière was well satisfied. On July 7 he wrote as follows:—

> These men are now being dressed in the cloaks from Vienna and the clothes brought back from Macerata.[97] They have been informed of the pay and allowances instituted by the law of 1852; they will get nothing more.
>
> I am going to set up a regular Conseil de Guerre[98] for them; they are all at the Lazaretto barracks where there are good prisons. They have as much devotion to the Holy Father as want of discipline.
>
> I have in no way committed myself to the ranks granted by Count Macdonald [sic] to Count Fitzgerald and the four captains. They are all Irish and have all served in Austria, the

---

these honest fellows, and that some pretense was made of keeping them, during their stay in Vienna, where the men were fed at a florin a day and that this justifies the complaints that they have raised since they arrived here."
—Letter of General de La Morcière written between July 13th and 17th, preserved in General Kanzler's collection.

A copy of this important letter was most kindly given me, together with many other documents, by the Marquis MacSwiney of Mashanaglass, who had received them from his friend the late Baron Kanzler, son of the General.

97 These uniforms rather disgusted the volunteers; they were not the Irish green and gold that they had looked forward to wearing; their own uniform was not ready, and was never served out to them. These were temporary, secondhand uniforms, some of them very dirty, and most of them far too small. They consisted of a blue tunic with wide tails, red breeches and white gaiters. The volunteers had looked for their own green and gold Zouave kit. Moreover some of the Tipperary giants of about six feet three high were invited to get into tunics and breeches made for men of about five feet four. Such difficulties as these did not exist of course for recruits in other units, which had a depot company and a clothing officer waiting to receive them.

98 Military council or court-martial; there were three such permanent courts in the State, but this one was never formed.

former as Captain and the other four as Lieutenants. They take a great deal of trouble about their men, and know their profession. I think we shall be able to confirm their ranks. We shall get here a very fine battalion of 500 men.

De La Moricière, G.-O.-C.[99]

There was however, a good deal of discontent over the broken conditions, and after de La Moricière's departure it found vent in a disturbance directed mainly against their own officers, whom they regarded as responsible for not seeing that the promises were fulfilled; Major Fitzgerald and Lieutenant O'Carroll and several foreign N.C.O's were hurt,[100] so, on the same evening, fifty-six malcontents were shipped back to Trieste to be followed soon afterwards by another twenty or so, after which matters went better.

There seems to be very little doubt that several at least of the malcontents were in reality agents paid to stir up unrest within the battalion.[101]

When writing about this disturbance, de La Moricière evidently thought there was a good deal to be said in favor of the malcontents, and entirely admitted that the promises had been broken. He wrote next day to the Minister of Arms:—

---

[99]   A.S.R. Com. Gen. tr. pont. 3154.
[100] Major Fitzgerald and Lieut. O'Carroll were probably not recognized as it was late, and they were in civilian clothes, and so were the Austrian N.C.O's. These officers were soon fit for duty again.
[101] "Des mauvais sujets qui sont meles aux Irlandais les excitent a ne pas rester tranquilles et menacent les officiers, les accusant d'etre la cause que l'on ne tient pas les promesses."—Letter of July 8th, from Col. Guerra to de La Moricifere, Archivio di Stato, Rome, Com. Gen. tr. pont, Busta 3154, Fascicolo 40.

De la Hoyde, who was serving as a lieutenant in Ancona, tells us in his letters that the tumult was headed by two men, one of whom seems to have been an Italian revolutionist, though he gave the name of Byrne; when his comrades tried to silence him, he shouted to the crowd in Italian for support, calling out that he was a patriot. The other was a man named Clark, who said he had served in a Highland regiment, and that before joining the Irish he had received £20 from the British Government to raise dissensions. After telling this story he was sent home.

Everyone was to some extent in fault, especially the Irish, and one of their officers who was not in uniform. I consider it certain that all this last detachment was enlisted by means of unreasonable promises and that they were kept at Vienna in order to be able to confer some military ranks.[102]

On the following day all was quiet—but there was one rather sad result of this disturbance, namely, that Major Fitzgerald resigned. It is said that, owing to his being an Austrian Staff Officer, his presence in Ancona had always been considered rather inadvisable, but he was a loss to the battalion: owing to the disturbance with the malcontents he considered his position compromised as commanding officer, but—so Lieutenant de la Hoyde tells us—he was so popular with most of the men that a crowd of them went to see him embark and one or two of them quite broke down at the last moment. So he returned to the Austrian army, not the first nor the last "exile" by many thousands, whose career has been impaired through trying to serve the land in which he was born.

After the above episode, matters went more smoothly; although the arrival of so large a number of recruits at the Lazzareto barracks occasioned at first a little confusion and consequent hardship for the newcomers, they seem to have settled down almost immediately to the daily routine of soldiering. And it was an attractive place to which they had come: a curve of hills jutting out into the blue Adriatic; in the hollow of the curve, a harbor and mole, where the sun blazes down on to the rippling waves; houses to the water's edge, a town built on cliff sides, as steep as a roof, commanded by the massive forts and bastion above, and occupied then by a garrison of about three thousand soldiers of various Catholic

---

[102] Teleg. July 5th—Ib. 3155 Fasc 45.

nationalities; hundreds of bright uniforms moving about the streets. This is the picture of Ancona in 1860.

In that small town the Irish volunteers drilled and sweltered for over two months on the barrack square. It was lucky for them that they had four regular company commanders, who, according to de La Moricière, were good and painstaking soldiers: they were Captain O'Mahony, born in County Cork, "a brilliant officer of the Austrian Uhlans," only lately returned from the slaughter of Magenta and Solferino; Count Russell of Killough, who had been for some years a Papal officer and had won' a decoration: Baron Guttemberg, Bavarian by birth, but an Austrian officer; and Captain O'Carroll, previously of the 18th Royal Irish. Under them training was almost continuous; having so many recruits in their ranks, the Irish officers were obliged to keep them at work most of the day: from four in the morning to seven o'clock they had drill; breakfast at eight o'clock; instruction in rooms from ten to eleven; dinner at midday; rest from one o'clock to three during the heat; drill from five to seven, all in barracks at eight, and silence at ten.

With their comrades-in-arms, the Irish, though raw recruits speaking no language but their own, did not take long to establish friendly relations. They were soon on good terms with the Swiss,[103] and with the Austrians[104] of whose soldierly qualities Lieutenant de la Hoyde speaks with the greatest admiration; and the French, of course, were always regarded as friends; the Italian Papalini alone seem to have remained rather aloof from the Irish, and indeed from all foreigners. After the first

---

103   "With the Swiss we agree well, but the men cannot talk the language."—Letter of Lieut, de la Hoyde, July 24th, 1860.
104   When at the beginning of September, an erroneous report reached Ancona that the Irish officers in Perugia, Captain Blackney and Lieut. Luther, had been killed in a revolt, Mr. de la Hoyde wrote: "We cannot, I think, spare troops from here, but doubtless when the Austrian reinforcement comes, they will be avenged, for the Austrians will be as vengeful for us as for themselves." And one of the veterans, Michael Fallon, by name, told me that "the Austrians were good comrades." Others confirmed this opinion.

The Half-Battalion in Ancona

three or four weeks the men lived comfortably enough; true, their pay was only three bajocchi, nominally a penny half-penny, per day, but it was equal to almost twice that sum in actual purchasing value, and their rations were good—though few of the volunteers liked the macaroni: a large brown loaf and coffee in the morning; for dinner, half a pound of meat, with rice, macaroni, and potatoes, and a good soup; in the evening, rice soup, coffee, or tea.

The only disagreeable feature seems to have been the hostility which many of the poorer classes in the town evinced towards the Papal troops of all nationalities; in the evening no soldier went out alone, and no officer without his arms. There were, too, several definite attempts made to discredit or corrupt the newcomers: firstly, the attempt of Byrne and Clark already described; then, a week or two later, several of the men who had visited an English ship said that they had been offered £4 a month to desert and serve as stokers, and on yet another occasion, the barracks were deluged with copies of a manifesto urging the men to abandon the Papal service. But all of these efforts failed. Later on, indeed, two of the men went off on board an English ship—but this is the only recorded instance of desertion among the whole of the Irish contingent, although it was extraordinarily prevalent among the other corps.

Under this steady routine, the companies soon began to improve: from the moment of their arrival General de La Moricière himself took a great interest in them, for in a letter dated July 24, Lieutenant de la Hoyde says: "Our sergeants, etc., got arms and were reviewed Friday last by La Moricière. He took the men only [through] 'Shoulder-arms,' 'Present,' 'Load,' and 'Charge a la Bayonette.' He seems to expect a scrimmage soon. He must take great interest in us, as he has reviewed us four times already, a thing he never does with the other troops till organized."[105]

The weather being hot throughout the months of July and August, it was, of course, impossible to do any active work except in the mornings and evenings, but the men showed excellent keenness. By August 5, when some of them had not yet been there a month, we hear that:

> Upwards of a hundred have got arms and turn out every day for target practice. They are very willing, and often drill in their room without orders...We are in first-rate order; only an occasional grumble at mistakes in mess, etc., which are unavoidable in a young corps, where none of us understand the language; but nothing to talk of.[106]

Here again—nearly seventy years later—we discover that Lieutenant de la Hoyde's opinion was shared by the General. On the same date, August 5, de La Moricière said in a telegram to Pimodan:

> The Irish in Ancona are going on perfectly. They are firing at the targets, and look very fine (ont

---

[105] That young de la Hoyde was right we know from de La Moricière's own letters and telegrams. Later on, when, during the hot weather in August, there was a certain amount of unrest throughout the garrison at Ancona, first among the Swiss and Austrians, and later among the Irish companies, de La Moricière wrote (August 15th) to Colonel Guerra, the officer in command: "Take all the trouble that you can, my dear Colonel, about these four companies; with them one requires patience, give and take, and an inflexible firmness in serious matters." On this occasion he sanctioned the appointment of another Second Lieutenant mainly as a disciplinarian, Mr. O'Connell, who won a decoration in 1861. O'Connell's appointment was partly due to a curious mistake; he had proved himself a good N.C.O., and his captain, O'Mahony, recommended him among others for promotion, observing at the same time that he was a nephew of the Liberator. De La Moricière at once exclaimed that any nephew of the Liberator must be given a commission, and two days later he appeared in the gazette. As a matter of fact this relationship was a mistake on the part of O'Mahony.

[106] De la Hoyde's letter.

très belle mine).[107]

On August 4 Baron Guttemberg, and on August 7 their other three company-commanders, Lieutenant O'Mahony, Lieutenant O'Carroll, and Count Russell of Killough were gazetted captains, and on August 8 de La Moricière said in a letter to Pimodan:—

> The 600 muskets of Jeannerat's battalion can, as you say, be distributed to the Irish [at Spoleto]. I thought they were already armed, like the Irish here [in Ancona] who already make fairly good practice firing at the targets.[108]

On August 28 the four companies in Ancona, after only seven weeks training, were considered fit to take their share of garrison duty together with the other battalions.[109]

---

107   De La M.'s dispatch, Aug. 5th—Ib. 3155.
108   De La M. to Pimodan. Aug. 8th—Ib. 3154.
109   Col. Guerra to de La M.—Ib. 3154. Fasc. 40.

CHAPTER VII.

Major O'Reilly's Work at Spoleto.

THE inevitable haste with which the battalion was formed is best proved by simply quoting the chief dates connected with it.

On June 12th, the order was issued by Cardinal de Mérode for the formation of battalion.[110]

On June 29, the cadres (establishment) of the battalion were published in the *Gazette*', this marks its first definite existence, but only four companies were planned in detail and only six officers were named.[111]

In the beginning of July, the scattered southern detachments began to assemble at Spoleto.

On July 10, it was ordered that each company should be 140 strong, including officers.

On July 24, 119 men at Spoleto who did not mean to serve took their departure, and on August 14 the last malcontents were at length allowed to go. This left about 580 stalwarts at Spoleto who at once began to make good progress; so did the 450 at Ancona.

On August 31, the four company commanders in Spoleto were gazetted captains.

On September 11, the war began.

---

110 A.S.R. M. A. Aff. Spec. 1169.
111 A.S.R. Com. Gen. tr. pent. 3155.

At Spoleto, therefore, when the invasion began, the volunteers had only had about six or seven weeks of regular uninterrupted work, and during that time, the officers had been shifted about: the captains had been gazetted only eleven days before, and the subalterns dated their existing rank from about three weeks earlier. In fact the half battalion was called upon to fight long before it had had time to learn its work.

No one, I think, not even a person of the strongest anti-Papal prejudices, could have avoided feeling some sympathy for Major O'Reilly during those months of July and August 1860. He had made the most unselfish sacrifice of position, popularity, and money, to say nothing of this daily personal work and the risk of losing his life for the sake of the cause which he considered that of his religion and his nationality.

The task before him was extremely difficult and discouraging: he had set out with high hopes that in spite of the suddenness of the call and the inevitable resulting confusion, he might help to produce another Irish Brigade which, in defending the Papal cause would, as in the eighteenth century, redeem the honor of Celtic Ireland. But, so far, his only reward had been—failure. The men were undisciplined—some of them out of hand after a month or more without authority; the officers, willing, but entirely untrained; and all around them were enemies. It was by now an almost herculean task to evolve an ordered soldierly unit out of this hurriedly-collected agglomeration of men, and he must have realized that war was very near. Yet, in spite of all these difficulties he still hoped to re-unite his thousand volunteers into one single unit, and so to have a strong and efficient Irish battalion ready for action when the fighting began.

With this end in view he seems to have evolved four main lines of action. Firstly, having got rid of the malcontents, to spread among the remainder an understanding of what military discipline really means; secondly, to train officers and men in drill and musketry; thirdly, to get muskets served out to them, and, at the same time, to raise a fund in Ireland for the purchase of rifled carbines in the place of the worn-out Papal smooth-bores; fourthly, as soon as matters at Spoleto should be working satisfactorily, to bring down the companies from Ancona, and reunite the Irish into one single unit. By these means he would get a fine battalion, over a thousand strong, and one that would arouse sympathy for the country from which it came.

At the beginning of July, the difficulties of the situation seemed insurmountable. O'Reilly was at Spoleto, which was to be the depot town of the battalion, working against time to get the place in order. But nothing was ready: the two hundred men with him, many of them[112] in civilian clothes, were sleeping in churches, lying on a scanty supply of damp straw; at the same time one hundred more were arriving from Rome where many others were already awaiting their turn to start. Meanwhile, up north, fresh batches kept landing at the port of Ancona. On July 7 (only nine weeks before the outbreak of war) O'Reilly was obliged to write to de La Moricière, asking him to prevent the latest-landed batches from leaving Ancona until he telegraphed for them, because the citadel at Spoleto could not be ready for another fifteen days.[113] This meant dividing up the battalion, for he was never again able to get those who remained at Ancona. It was a great pity, but it could not be helped.

Meanwhile, at Spoleto, everyone was uncomfortable. One of the survivors has spoken to me of this period, with some warmth: he said that many of his comrades were miserable because their pay was so small that they could not afford to buy more bread—the Papal ration was only fourteen ounces[114]—

---

112   O'Reilly's letter of July 7th.
113   *Ibid.*
114   De Mérode's letter, July 8th, *re* bread, and difficulty of language.

Major O'Reilly's Work at Spoleto

and, more. especially, because they had to go without tobacco. Nevertheless, in his letter, O'Reilly speaks of their satisfactory behavior.

In Rome, meanwhile, the news was that an English stranger had found his way among the Irish there, with offers of passports, money, and other inducements to desert, but the recruits fell upon him and would have made an end of him but for the interference of their Adjutant,[115] Lieutenant Howley. Another individual of the same type presented himself at Spoleto on July 10, but his presence was. reported to O'Reilly.[116]

In these circumstances it was, perhaps, natural, that the smallness of their pay should have become a grievance: they had been promised a shilling a day, and an enlistment bounty, neither of which was forthcoming: the bounty was paid later.

It was during this month of July, too, that the question of the malcontents came to a head. As already stated, about 210 of the volunteers refused to enlist, mainly because of the broken promises, but also because it involved sacrificing the four best years of their life to so very unhappy an existence. Pimodan tried to carry off the matter with a high hand by sentencing the men to cells, but this proved an absolute failure; it simply resulted in the prisoners' refusing to put on uniform. Finally, on July 25, 119 of them were allowed to return to Ireland, and, in the middle of August, about 44 more followed.[117] From that date a new era began.

The best opinion that I have been able to get about the conduct of these men is that of the late Mr. Crean, then serving as a subaltern at Spoleto. His views are especially valuable because he himself was very successful during this campaign; when the fighting began he won the highest decoration possible

---

[115] The Adjutant's letter of July 9th: Also that of the Marchese: Zappi, Colonel at the Papal G.H.Q., in Rome.
[116] O'Reilly's letter of July 11th.
[117] Pimodan's telegrams of July 24th; July 25th (second tel.) and Aug. 13th. A.S.R. Com. Gen. tr. pont. 3154 Fasc. 40. Also teleg. Aug. 21st (3152 Fasc. 5.)

for a subaltern and remained to his dying day an enthusiastic supporter of the Papal cause. But, to my surprise, he upheld to the end of his life that the men who went home had a very genuine grievance, and that they would not have insisted on doing so if the authorities had dealt more judiciously with them; he spoke of the unfulfilled promises, but more especially he blamed the government officials and contractors who, he said, were quite indifferent to the hardships suffered by the soldiers; so much so, that he always believed that there were revolutionists amongst them. In this last view he was right, for—besides the "ignoble speculation" already unveiled—on the very day before war broke out, General de La Moricière had two important commissariat officials arrested for treason.[118]

As a matter of fact, there certainly was something very peculiar in the direction of the stores during these months: the battalion could get nothing, although most of the things existed. So hundreds of men were going without cloaks, belts, shirts,' shoes, underclothes, collars, etc., etc.; in fact, they were short of the commonest necessaries, quite apart from articles of kit, such as haversacks, caps, etc., etc. As yet, however, the matter had not come to a head; but it was destined to do so, as will presently appear.

But, for Major O'Reilly, the most unfortunate feature of the situation was that, owing to the opposition of the malcontents, his immediate commanding officer, General Pimodan, was now in a state of permanent irritation with the battalion. It was extremely natural: one must realize that—as often during the nineteenth century—some of the Irish were their own worst enemies, and that there was probably no better soldier nor braver man in the Papal army than General the Marquis de Pimodan; but he was an officer of the old type, accustomed to an army in which the officers were nobles and

---
[118]   Telegrams of Aug. 4th, Sept. 10th and 11th—Ib. 3155.

the men conscripts, not volunteers. He gave orders rather as he might have given them to a pack of untrained hounds, with somewhat similar results, and did not realize that such hounds are brave and true-hearted, but that they need the man who understands them. This was a lesson which the Papal authorities were destined to learn within the next month or two. As war came nearer and nearer, they were very soon to find that throughout their army many of their rank and file had little heart in the cause, and that in some battalions about half the officers were politically[119] disaffected. In the Irish battalion this difficulty did not exist—as had been fully realized from the first by General de La Moricière.

And so, during the whole month of July, while O'Reilly and his officers were working under the most disheartening difficulties, Pimodan kept sending adverse reports to the General—who, however, does not seem to have paid much attention to them.

Yet one success had been achieved by Major O'Reilly, namely, that his whole-hearted efforts had won him the personal regard of Pimodan; about him, the Brigadier always reported well, and even in his most adverse letter he had written as follows:—

> Major O'Reilly seems to me capable of carrying out the work of his rank, and so far I have always found him sensible and full of foresight.[120]

And on his part O'Reilly fully realized the splendid qualities of his commanding officer: it was only two months later, when

---

119   On August 29th de La Moricière telegraphed to de Mérode saying that in the Italian regiments there were a large number of officers who did not want to serve in the coming war. He was sending six to Rome, and five had been disposed of elsewhere.
Both General Pimodan and General Schmidt had reported this to him, and now Colonel Lazzarini, hitherto incredulous, had formally confirmed their report. A.S.R. Com. Gen. tr. pont. 3154.
120   Pimodan. July 16th Ib. 3152.

he parted with him at Foligno, before the battle of Castelfidardo, where Pimodan was killed, that O'Reilly wrote in his diary: "This was the last that I ever saw of Georges Pimodan, the bravest soldier who ever lived."

MAJOR MYLES O'REILLY.

From a photograph in possesion of his daughter.

(Scan curtesy of Ave Maria Canizaro Library)

CHAPTER VIII.

Improvement.

So far, then, the course of the Spoleto companies had certainly not run smooth, but there were some consoling features in the situation: firstly, that its difficulties were due partly to the presence of a limited number of unsatisfactory individuals, but mainly, to bad organization; and that both these defects could be remedied. It might be added that, in many of the Papal battalions, there were defects far more serious than either of these two, which were in reality almost inevitable results of such a hurried enlistment.

That these defects were remediable was quickly proved as soon as the malcontent element had departed. There remained some 580 men at Spoleto, some of them still without uniforms, but all genuine enthusiasts, ready to work or to die if necessary. And towards the middle of August Pimodan left Spoleto to go on maneuvers, so O'Reilly was able to push on the training of his four companies with a free hand.[121]

With the end in view of producing a united Irish battalion, he still worked on his four original lines, which were: to discipline the men; to drill them; to get muskets served out to them while he tried to arrange in Ireland for private purchase of rifled carbines; and finally, to re-unite his volunteers by

---

[121]   *v.* Pimodan's telegram. Aug. 18th, 19th, etc.—A.S.R. Com. Gen. tr. pont. 3155.

bringing down the four companies from Ancona. After that, he might be able to get them all their full kit, and could hope to have a fine Irish battalion, fit for honorable service when the war began.

His first two purposes, the disciplining and drilling of the men, began to fulfil themselves almost at once of their own accord. But the last two he was never able to achieve before the outbreak of war, and in that fact lay the tragedy of the situation.

He kept his men hard at work all day, but for all that the veterans retained happy recollections of these last weeks before the war; the following is a description written to me by one of them:—

> Under the able command and sharp discipline of Major O'Reilly, we set to work at drill and other military duties in right good earnest; and so eager were all for the work—drill was almost continual—that in a short time we made wonderful progress.[122]

O'Reilly, of course, was very short of trained officers, though his four company commanders had had a little previous experience of military work. They were: Captain Blackney, a country gentleman from County Carlow, previously in the Militia, and shortly to win the Order of Cavaliere of the Ordine Piano for his conduct at Perugia, where he was wounded; Captain Coppinger, who had served in the British army and was also to win the Ordine Piano during the fight at Spoleto, where he too was wounded; Captain Boschan, of the Austrian service, a regular officer and popular with the men in spite of his foreign accent; and fourthly, Captain Kirwan, who apparently had served in the militia. But among the subalterns, unfortunately there were very few who had had any previous

---

122   Letter from Mr. M. Crean.

experience of soldiering,[123] though, of course, there was plenty of gallantry among them, and when the fighting began many of these untrained lads did astonishingly well. Lieutenant Crean was created Cavaliere of the Ordine Piano—a great honor for a subaltern—for his conduct during the defense of the gate of Spoleto where he was wounded; and Messrs. Stafford and Cronin were decorated with the Ordine S. Gregorio, which D'Arcy also won for gallantry in leading his half-company at Castelfidardo, although he was then only a boy of sixteen; in later years he rose to be a captain in the Papal Zouaves, and served in Mexico on the staff of the ill-fated Emperor Maximilian.

O'Reilly seems to have had some good N.C.O's, though perhaps not enough of them; and, above all, a good sergeant-major, Mulhall, who had previously been in one of the British regiments of Lancers and subsequently served with distinction through the American Civil War, rising to the rank of Colonel or General.[124] But it will be admitted that these companies were still at a most elementary stage; and the storm was coming very near them.

Still they remained cheerful, as will be gathered from the following brief description of his comrades by Mr. Crean:—

> After a short time things toned down into a course of pleasant routine. The influence of drill and discipline was becoming felt. The arrangements for the comfort of the men and their officers had become greatly improved, and in

---

[123] I can only find definite proof of previous military experience in the case of six officers at Spoleto: Captains Coppinger (British Army), Blackney (militia), Boschan (Austria); and subalterns Howley (10th Hussars), Carey (militia); but there may have been one or two more. Major O'Reilly himself had served only as a militia captain; and according to some accounts Captain Kirwan was an ex-militia officer.

[124] According to my informant, Mr. Bergin, he left the army with the rank of General.

point of fact there was not the slightest sign of the hardships of military life. We had an excellent mess at Spoleto, a good cook, an efficient mess-man, and I can only look back to that time with happiest recollections.

Some of our officers had served in the army at home. Captain Carroll[125] had been in the 18th Royal Irish; Howley, in the 10th Hussars; Coppinger had also sensed in a British regiment. In 1861 he joined the United States army and rose to a distinguished position. For many years after the war he was regarded as one of their ablest generals and retired only a few years ago.

Concerning the others we might add that Lieutenants Luther, Stafford, and Cronin also served during the American Civil War as officers in the Federal Army, the two former with some distinction. About Luther, one of his company who was with him during the fighting in Perugia, told me that he was the bravest man whom he had ever seen.

With one or two exceptions, these officers were all recruits, but they were beginning to have some knowledge of their profession. In a letter of August 24 to de La Moricière, O'Reilly told him that he had 600 men in Spoleto and that he was teaching them the use of arms.[126] He had got a Papal sergeant called Bell, who spoke several languages, to put all the officers and N.C.O's through the regular Italian course, after which he had set them to instruct a battalion-nucleus consisting of 160

---

125   This seems to be a slip on the part of Mr. Crean: Captain O'Carroll was in Ancona, not Spoleto.
126   "Le maniement des armes"—I believe this includes both the manual and also the firing positions. Major O'Reilly speaks of having had to go through this course himself, which shows that even those who had already learned the British system at home were obliged to start afresh; even an officer such as Major O'Reilly who was both knowledgeable and smart on parade, as we know from Pimodan's reports, and from the Irish veterans, was obliged to do so.

men specially selected by forties from each of his four companies. In his letter he added that the volunteers were now doing their share of garrison duty.

At the same time, he told the General that he had written to a cousin in Ireland about buying rifled carbines for the battalion, that the subscription had hung fire, but that his cousin said that a few written words from the general would enable him to buy them. This scheme of O'Reilly's was cut short by the war, but it was a matter of the greatest importance. Throughout the whole of the Papal army there was the keenest competition to get rifles instead of the smooth-bores. The rifled barrel was a new invention, and its importance was fully realized not only by the officers but also by the men. Pimodan had succeeded in securing them for all the battalions of his brigade, which was almost sure to be the first engaged,[127] but the other units including the Irish, were armed only with muskets and must therefore have felt at a great disadvantage as compared with the Piedmontese. The old Papal musket cannot have carried two hundred yards, whereas the rifles were sighted up to five hundred, and—so Pimodan said—would carry up to a thousand. One is not surprised, therefore, to hear that, when it was announced to the 2nd Austrian battalion that they were to have rifles instead of their smooth-bores, a "tremor of pleasure ran through all ranks" on parade.[128] The general then ordered that their cast-off muskets should be handed over to the Irish.

The improvement in discipline during these weeks was especially noticeable. Major O'Reilly in his diary speaks of receiving warm congratulations on this head, but the most convincing evidence is naturally that which comes from outside, and we include therefore the opinion of Count Alfonso

---

[127] One Irish company fought under Pimodan at Castelfidardo, but it was the only unit armed with smooth-bores of all those with him. De La Moricière's own brigade was entirely armed with smooth-bores, but with the exception of the Irish company, it was put in reserve.

[128] A.S.R. Com. Gen. tr. pont. 3154 V. also A.S.R. M.A. aff. ris. (1960).

Improvement

Visconti di Saliceto, the Piedmontese officer who was in command of the Spoleto prisoners at the end of the campaign, and spent several days marching them across Umbria, a mixed column, composed of Frenchmen, Germans Austrians, Swiss, Italians, and Irish. After dilating on the difficulty of keeping them in order, he says: "The most disciplined were a battalion composed of 500 Irish,"[129] and he was certainly in a position to know.

Meanwhile—failing the rifled-carbines—O'Reilly had been making every endeavor to get his men armed with smooth-bores, so that they might be in a condition to fight when the war began. In this aim the General strongly agreed with him; but, the General being away, a great many of the muskets never arrived. This was not his fault. As early as June 4, de La Moricière had allotted some arms for the Irish, then at Macerata, and when, on July 3, the 2nd Austrian battalion was served out with rifles, he ordered its cast-off muskets to be given to the Irish at Spoleto. On July 29 he even telegraphed to Pimodan his orders as to their method of instruction.

> I believe your Irish are armed. It will be enough if you simply teach them how to load and then take them down to the range. They will gradually learn the handling of their weapons. This method has been a success with those here [the Irish in Ancona].[130]

After such a definite order it is hard to understand why they were not given arms, but a week later Pimodan telegraphed, on August 6, that he was only awaiting orders to hand on to the Irish the 600 cast-off muskets of Jeannerat's Swiss Carbineers who had now received rifles.

---

129   V. Appendix. He was wrong as to their numbers, for there were only some three hundred and odd Irishmen in that column.
130   Telegram, July 29th—A.S.R. Com. Gen. tr. pont. 3155, Fasc. 45.

On August 8 de La Moricière replied from Ancona:

> I thought you had armed your Irish like those whom we have here, who already make fair practice on the range. Do not forget to instruct the officers first. Choose a knowledgeable officer from the Foreign Battalion to teach the officers the manual, just as I did here. When they know how to handle their arms they can teach it to the N.C.O's, and thence to the rank and file.[131]

De La Moricière believed in instructing the Irish rank and file through their own officers and N.C.O's rather than by appointing foreign instructors to the battalion, and this added to his popularity with them.

On the same evening, August 8, Pimodan wrote:

> Today I had the smooth-bore muskets left by the Carbineer battalion carried into the Rocca to arm the Irish when the time comes.[132]

The result of this last phrase, "when the time comes," was that the battalion was disappointed again.

Things had rather a way of "not getting done" in the Spoleto sub-division. Nevertheless, O'Reilly continued his training with unabated energy, and during the last three weeks of August there are no disturbances and no bad reports: at the end of that time, on August 29, General de La Moricière suddenly turned up to inspect the half-battalion.

There is no doubt that he went through everything with Major O'Reilly, but we have no record of what may have transpired between them, except the fact that some uncommonly "strong" letters and telegrams went flying in every direction.[133]

---

131   Letter. Aug. 8—Ib. 31544.
132   Ib. 3152.
133   Since the above was written Miss Edith O'Reilly, daughter of the Major, has supplied me, most kindly, with many interesting pieces of information

The Intendente Ferri received the first telegram—though, as often happens on such occasions, he was probably not the man who was really to be blamed. But this telegram is in itself a revelation of the conditions in which the volunteers had been serving.

> Tell His Excellency the Minister that the Irish at Spoleto are going on well but that they are short of—200 muskets—408 pouches—492 haversacks—100 cloaks—185 vests—75 caps—408 belts—300 shirts—300 pairs of drawers—300 pairs of shoes—400 collars—500 Mass-books—500 pairs of canvas gaiters—500 pairs of leather gaiters. Send everything that you can as soon as you are able and if possible some uniforms of the new model.—G.-O.-C.[134]

From this telegram (A.S.R. Com. gen. tr. pont Busta 3155 Fasc. 45) it will be seen that only thirteen days before the outbreak of war two-thirds of the Irish were still without belts, gaiters, haversacks, or pouches; and one-third without muskets, to say nothing of such everyday necessaries as shirts, underclothes, shoes, and Mass-books. And these were the

---

gleaned from her father. It seems that in after years he used to speak about this occasion. He said that the General inspected everything with him, and showed evident irritation at some of his discoveries. At the end, O'Reilly, who was proud of his companies' progress, asked the General to drill them on parade. At first de La Moricière refused: he could only give French words of command; but O'Reilly persisted, because he knew that "although the General was quick-tempered he was always just." Finally, therefore, they went out to the parade-ground where the four companies were drawn up, and de La Moricière ordered certain maneuvers, giving his commands rapidly in French. O'Reilly, however, in his capacity of C.O. translated each order as it was given, and the maneuvers were carried out quite creditably.

The result was, as is stated in the text, that the General reported favorably to the Minister for War; promoted four officers to be captains; selected one company to be "active service co." in Perugia and sent out some forcible letters to those responsible for the previous state of affairs.

134   De La M's telegram. Aug. 29—Ib. 3155 Fasc. 45.

conditions in which they had been living!

This telegram seems to breathe a certain suppressed indignation, but much stronger was the letter which the General wrote to Pimodan about the muskets: it was in fact surprisingly strong as being addressed to his second-in-command and close friend.

> To General Pimodan:
> Terni.
>
> I am astonished to find here 200 Irish without muskets, although it is now a month since I allowed them all to be armed: this deplorable result is the consequence of orders given by you to Colonel Lazzarini; in times like these it is hard to understand how muskets can be sent to Ancona without keeping a number sufficient to arm the men who are with you.
>
> Make all arrangements at once for supplying at Spoleto the 200 muskets required there. The fact that the Irish are unarmed compels me now to change all my plans: I need hardly say that if you have to get the muskets available at [Ancona?] or Pesaro, there must be no delay in sending them here. It is too bad that all the surplus muskets should have been sent to Ancona without keeping enough at Spoleto to arm the soldiers there.
>
> I must beg you, my dear General, to see that in future such mistakes do not occur in your command.
>
> I am, etc.,
>
> G.-O.-C.[135]

The answers to these letters are too long to insert here.

---

[135] A.S.R. Com. Gen. tr. pont. 3154.

Improvement

The Intendente forwarded some of the articles required, but not the haversacks, caps, cloaks, or sheaths for the dagas—all of which are indispensable for service in the field; nor could he send the green uniforms, as they were not yet far enough advanced.[136] Pimodan produced 180 muskets within a day or two, but they were the very last of the cast-offs, so that their striking power was uncertain, to say the least of it. Nevertheless they must have been among those used by the Irish during the fighting, although the men had not had time to learn their vagaries. So perhaps the most satisfactory result of de La Moricière's inspection was, in reality, the private letter which he wrote to the Minister of War, Cardinal de Mérode, expressing his own opinion about the battalion and its three weeks' work:

> Spoleto, August 29, 1860.
>
> Monseigneur:
>
> I have just made a detailed inspection here at Spoleto of the Battalion of St. Patrick, and it has made incontestable progress since I last saw it. To-morrow I am sending off to Perugia one of its companies,[137] 130 strong, to form part of the garrison in the citadel.
>
> Commandant O'Reilly has asked me to appoint captains to this right half-battalion, as it is over 500 strong and as yet has no officer of that rank.
>
> Consequently I have the honor to transmit to your Excellency the formal proposal herewith for four lieutenants of the battalion to be

---

136 Letter. Aug. 30 from Ferri—Ib. 3151
137 De La Moricière always stated this company at 130 men; but as a matter of fact, it marched officially 140 strong, and the Irish writers after the war stated it at 143 men; there were two officers with it, Captain Blackney and Lieut. Luther. *V.* telegram Aug. 31. Ib 3155.

promoted to the rank of captain.

I can still get a strong detachment from this battalion ready to march if required. I refer your Excellency to the notes herewith and beg to remain, etc.[138]

Two days later Lieutenants James Blackney, Martin Kirwan, John Coppinger, and August Boschan were promoted to be captains; this was only eleven days before the invasion began.

Thus, by the end of August, the Irish volunteers both at Spoleto and Ancona, though only half-trained, were counted upon to fight so far as their kit permitted. De La Moricière had sent one company to garrison Perugia, the principal danger spot in the state, and proposed to take another with his field-force if required. But this was not what they had hoped for: if they could only have been united, they would have formed a fine battalion, about 1,040 strong, full of enthusiasm, a true successor of the historic "brigades" of Ireland.

---

138 Letter. Aug. 29 Ib.

PART II.—CHAPTERS IX. TO XX.
THE CAMPAIGN OF CASTELFIDARDO, 1860. THE STORY OF THE IRISH VOLUNTEERS DURING THE FIGHTING.

CHAPTER IX.

THE PIEDMONTESE PREPARATIONS FOR INVASION. SEPTEMBER 1ST TO 10TH.

INVASION is now upon us, so we must leave the volunteers to their own devices and devote a chapter to the movements of their Piedmontese enemies during the last ten days before the declaration of war. During those last days Cavour was straining every nerve to raise rebellion in all parts of the Papal State before he actually sent his armies over the border. And in these efforts he showed extraordinary ability.

He constantly employed deception; there is now no doubt about that. But let us remember that this was. not for his own personal ends. It was to win unity and freedom for the whole Italian race; it was to realize a splendid patriotic ideal. And at that moment he was terribly hard pressed by the great powers. around him.

On August 31 Cavour wrote to Count Persano, Admiral of the Piedmontese fleet:

> With this end in view we have arranged as follows: a revolutionary movement will break out in those provinces [Umbria and the Marches] between the 8th and the 12th of September. Whether it is repressed or not we shall intervene.[139]

---
139   Admiral Persano's diary.

Already he had thought out all the details of his long-planned scheme for winning the war before it was actually declared; secret agents, arms, money, and propaganda had been sent into the Papal State to raise a general movement of insurrection; and presently they were to be followed by large bands of soldiers and trained officers. The chief aims in view were:

Firstly, political: to spread through Europe the idea that the Papal State was in rebellion, and was being repressed by foreign mercenary troops, and that the Piedmontese were coming in as deliverers and restorers of law and order.

Secondly, military: to draw away troops from the garrisons into wild pursuits of the irregular invaders so that the vital points should be unable to offer much resistance when the regular army of invasion made its entry.[140]

For the sake of clearness we will summarize their results at once.

These plans for fighting before the declaration of war were entirely successful. Some four or five days before declaring war, the Piedmontese sent bands of irregulars over the border to raise rebellion and to draw out some of the Papal troops in vain pursuits, and thus weaken the main body. In the north the irregulars seized Fossombrone; and two Papal columns, one about 800 strong under Kanzler and the other of about 1,000 men under Von Vogelsang, were at once sent out against them from Ancona. Similarly further south a body of irregulars under Masi was sent from over the Tuscan border

---

140   Incidentally it was intended to offer two lire (one and eight pence) a day to any Papal subject willing to fight against the Pope. This was munificent pay, being far above the agricultural wages, and far above the three halfpence a day doled out to the Papal soldier. It was hoped also to corrupt the Papal army as far as possible by bribes or other means; to organize a spy service, mine bridges, occupy important points, etc., At the same time the press propaganda rose to white heat; all these methods, it must be remembered, are considered perfectly legitimate in "the noble art" of war.

The Piedmontese Preparations for Invasion
September 1st to 10th

to seize the Papal town of Orvieto. This at once drew out the Papal C.O., General Schmidt, from Perugia with 1,250 men, in a vain pursuit of over thirty miles.

Meanwhile the Papal main body, now only about 6,500 strong, was at Foligno, Spoleto, and Terni, under command of de La Moricière.

And no sooner were the small Papal columns well scattered, than the Piedmontese armies rushed over the border to cut them off from Ancona and from Perugia respectively.

In the northern area Kanzler only just escaped after an ably-managed rearguard action at Sant-Angelo, and in the southern district Schmidt was captured in Perugia.

But we must describe the pre-war fighting in greater detail—for it concerns the Irish.

On September 6 the movement for stirring up rebellion began under the direction of the well-known Colonel Masi, an officer of the Piedmontese service, an ideal man for the work because he was not only a soldier, but a life-long enthusiast for the union of Italy; he had fought as a volunteer in 1848 and as an officer of Garibaldi during his defense of Rome in 1849; poet, orator, as well as fighting-man, he was at his best when leading irregulars.

On September 6—that is to say, five days before the declaration of war—Masi (so he tells us in his report) crossed the Pontifical frontier "by order of the Prince of Carignano"; on September 7, a body of Sienese, mostly armed, crossed it at another point, and various bands of rebels within the Papal state started on their secret night marches towards La Scarpetta, the rendezvous in the district of Orvieto where lay buried the muskets from over the border. By September 10 various small encounters had taken place with gendarmes and several villages had been occupied, and Masi found himself in command of a total force of over 800 men nominally Papal subjects in rebellion but of whom nearly 400 came from outside the state, including over 200 in the uniform of the Tuscan national guard. That night he attempted to escalade the rock of Orvieto but without success; on the following day, however, September 11, the day on which the war was declared, this town with its impregnable fortress[141] was surrendered to him without a fight at the wish of the Papal Delegate in authority there, because he and most of the citizens wanted to avoid bloodshed. And not long after this exploit Masi's battalion was taken onto the strength of the Piedmontese army under the name of the Cacciatori del Tevere, or Chasseurs of the Tiber, and fought under General Fanti's orders throughout the campaign.

On September 10 General Schmidt received orders to move in pursuit of Masi, so he started from Perugia with 1,250 men and two guns, leaving only about 400 in garrison. Thus by the time that war was declared the Piedmontese had already achieved their first object of publishing to Europe that the Papal state was in rebellion; and also—which was far more important—their second object of drawing away troops from the Papal garrisons in vain pursuits after raiders. Schmidt's was not the only instance; similar events were taking place further north at Urbino in the province of the Marches—but these do not concern the story of the Irish battalion. At the same time, in justice to the Papal Government, one must add another fact in its favor, namely that the attempt to raise rebellion in Umbria had proved a complete failure; even at one-and-eightpence a day the committees could hardly get any volunteers; and even in the more northern province of the Marches their success was not very much greater.[142]

---

141   Impregnable, that is, to Masi's force which had no artillery. It was garrisoned by no good Austrian Bersaglieri, and 28 Italian gendarmes under a young officer called Captain du Nord. They were too few to hold the town, but they might have held the fortress for weeks.

142   There were several small risings, supported and largely initiated from outside; but Cialdini telegraphed to Fanti that these would die out unless the invasion began as soon as possible.

The Piedmontese Preparations for Invasion
September 1st to 10th

And now let us in turn be just to the Piedmontese: when they accused the foreign troops of being oppressors and plunderers they were wrong,[143] but when they claimed that the Pope relied on foreigners[144] to support his government and thus to prevent the making of the new Italian nation, there was a great deal of truth in what they said: but for the presence of these foreign battalions, the active, enthusiastic, united-Italy movement would have carried the day over the heads of a more or less apathetic people. It was true that at least one-third of the Papal army consisted of genuine Italian Papalini born in the state, but in the autumn of 1860 there were very few of these native troops who thought the Papal rule worth fighting for—as became perfectly clear during the war. Thus, if left to themselves, the inhabitants of the Pope's dominions would have joined the new united Italy. What else could they do? The whole situation had changed during the previous year, and especially during the last six months. First Parma and Modena had united themselves to Piedmont; then Tuscany had followed their example, and Victor Emmanuel's capital had been transferred from Turin to the more southern city of Florence; finally, Sicily had risen, and now even Naples seemed to be on the verge of joining the movement. So one may fairly conclude that, but for the foreign defenders of the

---

[143] As a matter of fact there were probably more acts of violence attributable to the Italian gendarmes (police) than to any foreign corps, but they were not reported with the same zest.

[144] Ever since his return in 1849, Pius IX had been obliged to rely on Austrian troops to support his authority; but this was partly due to the spirit of rebellion in the towns of Romagna and the northern districts, which in 1860 were no longer under his rule. Undoubtedly any government which relies on foreign troops is, *ipso facto,* self-condemned; but one regrets that Pius IX of all people should have this stigma attached to his name, because he had begun his reign full of ambition to be at once a Liberal Pope and a good Italian; he had done everything possible to achieve this end, even appointing noted revolutionists such as Rossi and General Zucchi to be his prime minister and minister of war; and men like Galletti, the republican, to other cabinet posts. It was only when Rossi was murdered, Pius himself driven out, and a republic set up in Rome that he saw himself compelled to resort to repression or lose the Church's inheritance.

ancient rights of the Pope, his subjects would have followed the same course as the other Italic peoples, simply because there was nothing else left for them to do.

But they would not have done so with anything like universal enthusiasm except in certain of the towns, such as Perugia, Foligno, Macerata, and some others. The peasantry would have been apathetic, or in Umbria even reluctant; but the active political part of the population would have carried the day—as it still carries the day. And the chief influence in its favor was surely the wonderful story of Garibaldi in Sicily: he supplied them with a new and glorious fighting tradition of which any people in the world might be proud.

Thus when war was declared, on September 11, 1860, the Papal army found itself in an impossible position: it was now the defender of a cause which had hardly any vitality left. The Pontifical officers and men were about to fight simply for the right of the Pope to rule over a small international state in order to preserve his independence and his dignity as ruler of the international Church. To the Church party, and especially to Pius IX himself, the preservation of his right was a sacred charge; these territories had been bestowed on him by a Higher Power and it was his bounden duty to hand them on intact to his successor. But they were face to face with the sentiment of nationality—perhaps the most heart-felt creed of the nineteenth century, and the only one for which almost every man and woman was willing to offer his or her life—and also with the conviction that until Italy were united she would never be able to develop in safety.

The Papal army, then, was about to defend a right which was historically just, but, in September 1860, had become a practical impossibility. Nevertheless, as long as the government at Rome said Fight, it was the duty of its army to fight. This strange situation had affected the morale of the whole force, and had worked on the different nationalities and temperaments with the most varying, even opposite, results.

Among the officers and men, as also among the clergy, there seem to have been plenty who realized the futility of the position—hence surrenders such as that of the rocca of Orvieto—but the most genuine or the most patriotic among the soldiers believed that they must offer their lives in defense of their own oath of allegiance or their country's honor.[145]

It is now time to enquire once again how the Irish volunteers had fared amid all these pre-war movements.

On August 31, one of the four companies had been moved from Spoleto to Perugia, which is the chief frontier town in the west of the state; this company was not sufficiently equipped to fight in the field, but it was now sufficiently trained to fight as part of the garrison if necessary. It was the first to be made an "active service company," and consequently it was at once besieged[146] by volunteers begging to be transferred to it. De La Moricière had originally spoken of it as being 130 strong, but before starting it certainly consisted of 143 men and two officers, who marched away from the massive old oblong castle above Spoleto in the highest spirits, followed by loud cheers from the comrades remaining there. But the departure of this company must have been a rather serious loss to O'Reilly, especially as it took two good- officers and several valuable N.C.O's, namely: Captain Blackney, Lieut. Luther, Sergeant Hoey, who was a good drill-instructor, and one or two other sergeants; to say nothing of Corporals Synan and Allman, whose names are now well-known.

The town to which they were going, Perugia—known as Augusta Perusia in the days of the first Roman Emperor, and

---

145   Major O'Reilly says in his diary: We were fighting only for honor. He realized that no material success was possible.

146   "The moment the word got out that No. 1 Company had "been made 'active service co.,' the staff officers' quarters were besieged with crowds begging and arguing to be allowed to volunteer into No. 1, which they said they heard was ordered to the field."—*The Nation* (Dublin), 8th of September, 1860.

before that, already a historic Etruscan town—is an ancient city standing on the summit of a hill and surrounded by three or four miles of mediaeval walls, varied here and there by immense bastions, perhaps sixty or seventy feet high, and for several centuries Perugia had also been guarded by the huge castle-fortress erected by Pope Paul III at the south eastern end, and stretching halfway down the hillside. These old fortifications gave an impression of immense strength, but in reality they were no longer of any great value: the walls were not seriously defensible against artillery, even in 1860, and the old Paoline rocca had been pulled down by the revolutionists themselves some years before. It was only in process of being rebuilt; but still de La Moricière believed that "with 400 men of good soldierly quality" it could be held for several days, and that if more men were available its resistance could be prolonged by occupying the houses around; so when the fighting became imminent over 400 men were told off for the fortress, and General Schmidt, with 1,250 more, was left there to be employed as required.

The remarkable feature about Perugia—which then contained 18,000 inhabitants closely packed within its circle of walls—is the fact that it is a mediaeval town which has continued almost unaltered down to our own time. It consists of strongly-built houses between which the narrow streets wind up and down, at some points cut into broad stairways, often diving under old gothic arches, hardly ever straight, and very difficult for strangers to learn. It would have been an ideal town for street-fighting if the people had been willing to defend it: but the people were the most bitter revolutionists in the whole state and hated the Papal government.

It was in this quaint old place that the 145 Irishmen found themselves in the beginning of September, entirely unaware that all over the country around them the secret committees were at work and the bands of drilled men moving from

The Piedmontese Preparations for Invasion
September 1st to 10th

place to place every night. They themselves were comfortable enough, forming part of the regular garrison, together with a company of Italians 75 strong, a company of 120 Swiss,[147] and some Italian and Swiss details for commissariat work, etc., just about 400 men in all.

On Sunday, September 9, the Irish had their first ocular demonstration of the work that was secretly going on outside. For a vivid account of this and other episodes we are indebted to a letter signed "P.F.C." in the *Waterford Citizen,* reprinted in the Nation of October 13, 1860. These initials are those of a volunteer named Clooney, a man of no great education, but of splendid courage: he afterwards rose to be a captain in the American army, and was killed during the Civil War on the terrible day of Antietam, but has since been remembered by his brother officers as "the bravest of the brave." The following is the account he gives:—

> We... were stationed in one of the palaces—that of the Corso. Time wore on for a few days and we were in expectation of an attack, but it did not come off at this time. However, on the Sunday evening after, I was sitting with a friend on one of the walls of the fortress, speaking of poor old Ireland, when we were interrupted by the noise of persons on horseback flashing under the walls beneath us, and looking down we beheld

---

147   It is not very clear whether there was a company of Swiss there permanently or not; a Perugian correspondent to the *Nazione,* of September 7th, speaks of the three companies above-named, but Colonel Vigevano in the Italian official history of the campaign speaks of an Irish and an Italian company as being the regular garrison. When Schmidt took his whole brigade to pursue Masi, another Swiss battalion was sent to Perugia to take their place, but was recalled to join de La Moricière's field-force on September 11th, thus leaving very few troops in the citadel.
   V. M. G. Uff. Stor.
   For de La Moricière's statement of the garrison, V. A.S.R. Com. Gen. tr. pont. 3154. He states the Irish at 130, but two telegrams reported them at 140. They were, I think, 143 and two officers.

a mounted gendarme sweeping under the gates of the fort. He had not passed us, scarcely, when another appeared, his horse dripping with sweat, and himself in apparent consternation. After him followed another bearing on his horse's back the clothes of a wounded comrade. The three appeared at the gate of the fortress, and being recognized as Pontifical gendarmerie were admitted within the fort. They conveyed the information that u band of rebels from Tuscany which had been hovering about the frontier for some days had entered one of the villages by surprise, routed the gendarmerie, had taken a captain prisoner with four men, and murdered one of their comrades.

This was news of some importance to us, and we were ordered to be ready to turn out at a moment's notice as an advance on Perugia was expected. Of course we were ready in little or no time, but no attack was made by the supposed rebels...

It is not quite easy to decide where these gendarmes had come from; their own story would lead one to think that they had been part of the small detachment in Città della Pieve, the first village taken by Masi; but his account of that affair is that the gendarmes there all deserted at once, musket in hand, and joined his invading force; so it seems more likely that these were the remnants of another batch who, when coming out of church on September 8, had been ambushed and fired on by a band of twelve Perugians on the march for Masi's rendezvous at Chiusi: they had lost one killed, one wounded, and two prisoners, only one man escaping unhurt to tell the tale.

The receipt of this bad news roused the Papal troops to

action. On September 10 General Schmidt started from Perugia in pursuit of Masi. He took with him 1,250 men, half Italian and half Swiss, with two guns, thus leaving in Perugia a garrison of two or three companies: one Italian, one Irish, and possibly one Swiss, with detachments of artillery, gendarmes, and finanzieri (customs-men)—only about 400 men at most—to hold the fortress, which would be one of the first points attacked by the Piedmontese army.

On that same evening the Piedmontese ultimatum had been delivered in Rome, and their regular forces had crossed the frontier; in another day their right wing, 12,000 strong, would be only a little further distant from Perugia than was General Schmidt, now careering off on his wild-goose chase after Masi. But the Irish knew nothing of the declaration of war, nor did the commanding officers in Perugia, nor did General Schmidt himself; and so, as Clooney says, the days wore on.

CHAPTER X.

WAR DECLARED: THE STRATEGICAL PLANS ON EITHER SIDE.

AT this point it may be as well to give a very short sketch of the plans of campaign on either side, so that any reader may understand the reasons for the various movements.[148]

Quite briefly the situation was as follows: the Papal frontier stretches diagonally across Italy from north-east east to south-west. Upon the north-eastern Adriatic coast was the port of Ancona; down in the southern part of the State was Rome, and these were the only two places strong enough to stand a siege; Rome, in fact, was perfectly safe, being under French protection.

Between the two was the main body of de La Moricière's army, stationed at Foligno, Spoleto, and Temi; but it was fairly evident that when war began he would seek refuge in either one or the other of his two strong places, Rome or Ancona, because his army was entirely unequal in numbers, training, and equipment to that of the Piedmontese; it was practically certain that he would not attempt to meet his enemies in the open field.

De La Moricière could, of course, have retired southwards

---

148   *V*. Reports of de La Moricière, Fanti, Cialdini, Della, Rocca, etc. Books: The official history by Vigevano: and La Battaglia di Castelfidardo by Barbarich *V*. also Corvetto Corsi and other military writers.

on Rome, and have remained there in perfect safety, ready to harass the Piedmontese if they attempted to march past him in order to attack Naples, as was their intention. But this course would have involved abandoning Umbria and the Marches, in fact, most of the Papal state, and handing them over to the invaders without a blow. He preferred to march northward and occupy Ancona. It was fairly well fortified; it was so placed that from the shelter of its forts he could organize a series of attacks upon the invaders, perhaps threaten their communications, and in any case prevent their feeling secure within the occupied territories. And Ancona was a seaport on the Adriatic, that is to say, a town very much before the eyes of Europe; a gallant defense of Ancona lasting perhaps two months or more, and varied by rapid dashes out and swift blows at the besiegers—a type of warfare in which de La Moricière was a past master—might arouse the sympathy of Europe and the interference of the Catholic powers in favour of peace. It was a bold plan, but to de La Moricière it seemed better than a calm abandonment of the Papal provinces. "If I had failed for want of boldness," he said afterwards, "my old African friends would have cut me dead."

For the success of this plan, however, it was absolutely essential that his field force at Temi should be certain of having time to march northward and get safely into Ancona, before it could be cut off from that town by any invading force crossing the northern frontier to intercept it. This was the point of which de La Moricière had not made quite sufficiently certain and the result proved fatal to him. The campaign resolved itself into a race for Ancona; de La Moricière did a lightning march of six days northwards, but Cialdini, rushing down from Rimini and La Cattolica, succeeding in intercepting him at Castelfidardo. They fought: with his own African French troops de la Moricière would probably have got through, but with the untrained men at his disposal he failed to do so.

On the Piedmontese—or, one should in reality say, the North Italian—side, General Fanti, of course, realized de La Moricière's position, and foresaw that he would probably make for one of the two strong places. Fanti, therefore, divided his army of invasion into three: his left wing under General Cialdini was to cross the frontier at La Cattolica, right up in the north, with 17,000 men, and rush along the Adriatic shore to cut off de La Moricière from Ancona: his right wing was to march on Perugia, 12,000 strong, to take that town and make it a base for further operations against de La Moricière in case he concentrated his men for a fight around Spoleto or Temi instead of going northwards to Ancona; and between the right and the left wings was a central division over 7,000 strong marching along the mountain roads to Gubbio, to keep connection between the two wings and reinforce either if necessary. Fanti had thus about 38,000 men[149] in the field, if we include the irregulars of Masi and one or two other responsible leaders. De La Moricière had nominally about 21,000, but of these only 18,000 could be called mobile, and after deducting garrisons, he never could have got more than about 12,000 for service in the field, as against Fanti's 34,000 to 38,000; and most of these were poorly armed and—if the truth be told—without much spirit for the conflict.

Fanti's first object was to capture Perugia before Schmidt could get back there, and he took time by the forelock. He crossed the western frontier with Della Rocca's 12,000 men—his right wing—on September 10, the evening before war was declared: next morning he took the small town of Citta di Castello defended by only a few gendarmes, and pushed on rapidly towards Perugia.

Meanwhile Cialdini with the left wing, 17,000 strong, had crossed the frontier at its northernmost point, on the Adriatic.

---

149   The Italian official history says that only 34,000 regular troops were present on the day of the invasion. On paper there were several thousand more.

War Declared: The Strategical Plans on Either Side

And the central column, under General Cadorna,[150] was also in motion towards Gubbio.

Thus the invaders of the Papal state had crossed the frontier at three different points simultaneously.

From what has been already said, it will be evident that when, on September 11, the regular invasion began, it found the small Papal army completely unprepared. In Umbria, Schmidt was far from Perugia, in which so few troops remained that at first the Piedmontese thought they would be allowed to occupy it without resistance.[151] In the Marches, after a similar raid, the towns of Urbino and Pergola had been occupied by irregulars and rebels, and two columns of Papal troops under Colonels von Vogelsang and Kanzler were scouring the country in vain. In the center of the state was General de La Moricière with about 6,500 men round Terni, Spoleto, and Foligno. In short, his army, which should have been concentrated to meet the invading army, had been dispersed far and wide in pursuit of the irregular raiders, who had escaped.

---

150   The same who entered Rome by the Porta Pia in 1870. He was the father of General Cadorna who commanded the Italian army during the first part of the Great War.

151   Dispatch of General de Sonnaz, 12th September, 1860. U. S. 42, 12. Quoted in the Italian official history of the campaign, p. 207.

CHAPTER XI.

Position of the Irish Battalion at the Outbreak of War.

Meanwhile, we must return to our subject and ask how the volunteers had fared during these September days. The fact was, of course, that the war had come long before they were equipped for service in the field and before they were reunited. Major O'Reilly had continued his efforts, but without success. He had, it is true, got muskets, but there had not been time for the rifles to arrive.

His attempt to reunite the battalion had also failed, at all events, for the time being. He had been up to Ancona for several days to see the four companies there, and had at once written to the General a diplomatic letter asking for leave to bring the four Ancona companies down to Spoleto, on the plea of completing their training, but de La Moricière had refused. Still, one feels that they might have been reunited later on, but at that moment, with invasion so close, they were required for garrison at Ancona, where de La Moricière intended to make his chief stand.

From this condition of affairs there arose the rather heartbreaking result that when the invasion began they had to abandon all hope of fighting as an Irish battalion and winning glory for their country in the way of the old brigades. They were equipped only for garrison work, and consequently must expect to be divided up and scattered; one company at Perugia, two at Spoleto, four in Ancona, and only one marching with de La Moricière. That was to be their fate during the coming war: to be swamped amid the greater numbers of foreign troops. Their own green uniforms had not arrived, and the old ones in which they stood were similar to those of the foreigners, so that every external sign of identity was to be merged in the general mass: as far as outward appearances were concerned they would be indistinguishable from the others.

That was the fate which actually befell them. But, nevertheless the battalion was able to make good; and this was because it was the most exclusively national battalion in the Papal service, composed entirely of men of one nation. It contained practically no one but Irishmen, and consequently, though divided up, it remained alive and sentient in each of its scattered members, and their national esprit-de-corps pulled each of them through with honor in the end. And it is surely a matter for pride that although they were thus separated into four small units, each lost among the greater numbers of the regular battalions, nevertheless, these half-trained recruits never allowed themselves to be demoralized by the often contemptible example of the old soldiers around them, but at the end of the campaign, every single one of the four isolated Irish units obtained an honorable mention in the official report of its respective commanding -officer; although in the one instance he was a Swiss, in another an Austrian, and in the third a Frenchman—the fourth being General de La Moricière himself. And excepting the Franco-Belgians, our volunteers won more decorations than did any other battalion in the Papal army.

CHAPTER XII.

The Capture of Perugia.[152]

MEANWHILE, General Schmidt, after wasting a day or more rather unaccountably, had begun a forced march back to Perugia, hoping to get there before the Piedmontese could arrive—though he tells us that he did not know he had anything but bands of irregular raiders to deal with. In fact, on this south-western side of the state the movements of the two opposing forces developed into a race for Perugia, which ended in a dead-heat. At 7 a.m. on the morning of September 13, Schmidt arrived back there with his 1,250 men tired out after covering 27 miles in 14 hours, ending up with an all-night march. And he returned only to fall into the jaws of the wolf, for Fanti and Della Rocca made their appearance

---

152   This chapter was submitted to one of the veterans who had taken part in the fighting—namely, Brother Howlin of St. Patrick's Monastery, Mallow. I also had a long talk with another ex-soldier named MacCorry.

Clooney's letter also gives good Irish information, of which one finds some satisfactory confirmation in the Italian archives.

Official Reports: On Italian side those of Generals Fanti, De la Rocca, Di Sonnaz, di Savoiroux, and junior officers. Most of these are printed in Degli Azzi's book, *La Liberazione di Perugia.*

Official Reports on Papal side: De La Moncière; Schmidt (A.S.R. Com.-gen.-tr. pont. 3151), Lazzarini. And, in the Vatican Archives there is a private report by Schmidt.

There are, of course, very many other accounts of the fall of Perugia. (*V.* List of authorities for some of them. But the best are the official work of Vigevano (*La Campagna delle March e dell' Umbria*) and Degli Azzi's work (*La Liberazione di Perugia*).

about half-an-hour later with their 12,000 men comparatively fresh, and entered the town at once through a gate handed over to them by the inhabitants. The Piedmontese generals promptly sent forward a strong body of troops, headed by the 16th Bersaglieri and two guns, with orders to rush through the streets regardless of firing, and occupy the cathedral square; whilst various other battalions, instead of entering the town at once, inclined to the left, skirting the walls, and attacked the Santa Margherita gate, which they forced after half an hour's fighting against the Swiss, commanded at this point by General Schmidt in person. Having thus got a firm foothold within the place by two different entrances they proceeded to occupy positions all around its walls so as to prevent the Papal troops from making their escape.

There then resulted about three hours of rather confused fighting in the streets, of firing out of windows and of occasional bayonet struggles, during which some of the Italian Papal troops, though weary from their long march, showed excellent spirit, but on the other hand many of the Swiss surrendered with surprising ease; in all his battalions, however, General Schmidt tells us, he noticed that a good many of the officers were not with their company, and in several instances they refused point-blank to leave the fortress.[153] And in one building a batch of about 100 Swiss surrendered to the Piedmontese engineers without there being even one single casualty in either unit. On the Piedmontese side one may say that all the troops, and especially the Bersaglieri and the Grenadiers, showed fine dash, and enthusiasm worthy of better adversaries; but the Papal defense was very unequal in quality. Still it lasted until 10 a.m., when a white flag was raised by the Piedmontese and a truce agreed upon until 3 p.m. This truce was the beginning of the end; it was renewed

---

153   In the Vatican archives there is a document giving the names of these officers.

for an hour, and then was indefinitely prolonged. By that time the Papal troops had no thought of further resistance in the town—indeed a half-company of theirs had already laid down its arms and fraternized with the enemy—but some of them were still ready to hold the fortress. At 5.50, however, when the cannonade was renewed, they surrendered.

During the day's fighting the Piedmontese had lost seven killed and 56 wounded; the Papal troops, which were over 1,600 strong, had lost 37 dead and 60 wounded; among the dead were two officers, whose names appear to be Swiss.

From this brief sketch it will be seen that, from the very first, the Papalini had little chance of success: and also that some of them had very little heart in the fight, especially some of the officers. Schmidt said in his report that he surrendered because he could not get his troops to continue the fight; in many cases the officers were completely tired out and discouraged, while inside the fortress only three companies remained at their post to the end.

* * * * * *

Let us see how the 145 Irishmen had fared during this disheartening day.

The first point that we may note is that Captain Blackney was among the wounded, and that Lieutenant Luther, though not wounded, had certainly shown equally good courage; he was described to me by one of the veterans as "the bravest man I ever saw," and in all their communications they speak of him in similar strain. Of the N.C.O's, Corporal Allman had been killed and Corporal Synan wounded two or three times—he was afterwards created 'a Cavaliere of the Order of St. Sylvester for his services on this occasion; of the sergeants there is no mention, except a telegram which says that two of the Irish sergeants had been killed; but this message is unconfirmed, so its writer may have been referring to the Corporals

Allman and Synan, or he may have included one of them;[154] still, even so, it will be admitted that the company officers had duly paid their toll. Of the men we have no record, but we know that of Allman's twenty men alone at least six privates were hit as well as the two corporals above-mentioned: and, of course, there must have been an average number of casualties throughout the rest of the company, for it was engaged both in the fortress and in the streets.

The Irishmen had only been about ten days in Perugia when the invasion began, and during that brief period had been quartered in the Palazzo Donnini, which was near the fortress, and rather dominated it; the Papal officers evidently intended to hold these outside buildings so as to prolong the defense, according to de La Moricière's programme. During this short time the volunteers probably had very little leisure for exploring the city, as they must have been busy settling into barracks for the first day or two, and during the last two were hard at work. Meanwhile the enemy was coming nearer and nearer: the Italian commissariat officers were collecting supplies from a very unwilling population; and on the 13th the Irish were ordered to leave the Palazzo Donnini and sleep inside the fortress, each man lying out in the courtyard with his musket beside him; an attack was evidently expected, and they spent a good deal of the night in singing.

At 7 a.m. on the 14th, General Schmidt turned up with his tired-out brigade, and the Irish company was ordered to furnish two guards of twenty men each, one for the Archbishop's palace which looks on to the Piazza S. Lorenzo, that is to say, the cathedral square, in the center of the town, and the other, under Corporal Allman, for the S. Angelo gate, which is at the very farthest point possible from the fortress. To reach the S. Angelo gate it is necessary to go outside the inner ring of

---

154   I do not think that Sergeant Hoey was wounded; he certainly was not killed. But there seems to be a vague recollection that another sergeant was hit.

walls to the far end of a narrow suburb which juts out into the country; it is in fact over twenty-five minutes' walk from the fortress. According to the Italian official narrative there were various other small bodies of Irish as well as Italians sent out to occupy points in the direction of the Pesa and S. Antonio gates, and it adds that they fought among the other defenders there. The main body of the company remained in the fortress for the time being.

It was not long before both these guards were under fire. When the Piedmontese, headed by the 16th Bersaglieri with their two guns, dashed into the cathedral square, the first-named Irish guard, namely that in the Archbishop's palace, overlooking the square, was ordered to retire to the fortress. With some difficulty, therefore, these twenty men made their way through the up-and-down streets and archways and reached the castello apparently without casualties, though after nearly firing on some of their own Swiss allies by the way. They had got off lightly.

But for Allman's guard at the far-off S. Angelo gate, the situation was very different. With the name of Allman we come to one of the best-known episodes connected with the Battalion of St. Patrick, and it is also one of the best attested.[155]

---

[155] Apart from many casual accounts given me by veterans, or by their sons, we have the following authorities:—

Major O'Reilly in his speech at Wexford refers to it.

The letter from Clooney in the *Waterford Citizen,* October 13th, 1860, describing his experiences. It is very satisfactory to be able to confirm nearly all the names which Clooney mentions, by documents now in the Royal Archives of Italy.

Narrative sent me by Bro. Howlin, who was a friend of Allman.

Detailed account given me by Mr. Bergin, whose father was in the battalion, and a friend of Allman.

There are various slighter references to it, such as that in The O'Clery's book, *The Making of Italy;* and it has obtained a mention in the Italian official military history of the campaign. (M. G. Uff., Stor., p. 224.)

In the Papal documents, now preserved in the Royal Archives of Italy, we find various mentions of the men who were with Allman : Synan, Power, Murphy, and also Diamond are named; and we learn that Corporal Synan, who was second-in-command to Allman, was promoted sergeant, and made Cavaliere of the Order

Capture of Perugia; view from outside of the gate of Sant' Angelo, where Allman's party was posted.

(Scan curtesy of Ave Maria Canizaro Library)

---

of St. Sylvester in recognition of his services on this occasion. After the war he returned to Rome and joined the Company of St. Patrick then formed, but his wounds unfitted him for further service. In the rolls of the Company of St. Patrick (A.R.S. Busta, 1861), we find the following entry: Sergeant Synan (William) in Perugia on September 17th. Returned from Marseille to Rome on November 4th; received into this corps on the same day. Pay from Sept. 17th. On furlough for three months from November 26th.

Allman was in reality a young medical student from Cork who had interrupted his career and volunteered for the war from sheer enthusiasm for the cause. On this morning, the 14th September, at about 7.30, he and his nineteen companions reached the gate of St. Angelo, which is guarded by an old square red-brick tower, but they found that its doors had been stupidly locked against them. This, however, did not trouble them very much, as their orders were, if seriously attacked, to open fire and then retire towards the fortress.

When they had been in this outlying post for about three-quarters of an hour, they suddenly heard "heavy, sharp cross-firing" in several parts of the town, between them and their fortress, and they knew that the Piedmontese must have entered by one of the gates in the inner circle of walls. The big guns in the fort began to boom, which was proof that the enemy were in the center of the town, and consequently that they themselves were cut off from their main body; also that their guarding of the gate was no longer of any use. In another few minutes the firing came nearer and suddenly a body of soldiers considerably superior to them in number appeared at the top of a street in their rear. Clooney describes how the Irish hesitated to fire, being uncertain as to uniforms, but all doubt was soon ended by the Piedmontese letting fly a volley and then charging down on top of them: they returned the fire and then slipped away down a side street leading southwards to try and make their way back to the fort.

From that moment it was a running fight; they were obliged to keep a parallel course to that of the street along which the enemy were moving, so that every time they cleared a crossroad they received a volley: and soon they found themselves passing through "strange streets we had never seen." It was then that one of their number was badly hit in the leg and made prisoner, and soon afterwards Clooney was slightly wounded in the arm. A few moments later their way ran into a cross-street, and as they entered it they were fired on from either side at once, both ends being occupied by the Piedmontese. The nineteen Irishmen drew back and after a short consultation decided to try a joint rush. It was their only chance of getting back to the fortress. Having carefully formed them up behind him, Allman gave the word, "Now, boys!" and dashed out in front, followed by the rest. But the attempt was useless. He fell very soon, shot through the heart, and Corporal Synan, the next in command, received a bullet through his jaw; the third, a man named Power was brought to the ground with a bullet through his ankle; another boy, Magrath by name, was slightly wounded in the breast, and yet another was hit in the thigh; so, after emptying their muskets, they drew back behind cover. But at this moment we hear the story of what must have been a very gallant action: one of their number, a man named Murphy, seeing Power still wounded out in the open street, dashed out regardless of the bullets and carried him in and set him in safety with his back against the wall.[156]

The guard then split up into two parties: six of them dashed across the road and got under cover, but they were obliged to surrender not long afterwards; the other twelve turned back, and wheeling round through a small street, broke open the door of a house with the butt-ends of their muskets in hopes

---

[156] This action of Murphy's was told me by Mr. Bergin, whose father was one of the battalion, though not actually at Perugia, and Mr. Bergin told me that his father received the story from Power himself. I see that among the Irishmen who received the cross of St. Sylvester was a man named Murphy, and there does not seem to have been any other man of the name who had claim to such distinction. Mr. Bergin also told me that he was known as "mad Murphy," owing to his exploits, and that he was responsible for one of the rows at Macerata, as already mentioned. Power's wound was so serious that when taken to hospital the doctors wanted to amputate his foot, but he refused point-blank to have the operation, and his ankle afterwards made a complete recovery. He was a man fairly well known in Ireland, being the brother of a prominent Fenian.

There are several documents in the Archives in Rome giving further details about Power; and also several mentioning Synan. This, of course, together with Mr. Bergin's narrative, affords satisfactory confirmation of Clooney's letter. A.S.R. M.A. 1169. Compagnia S. Patrizio Matricola 1861.

that they might remain unobserved there until evening; but an hour later they found that they had been discovered and their house was surrounded by the Piedmontese.

Such moments bring out the natural leader: and in this instance it is Clooney who appears; tall, powerful, and fearless, as, later, on many a battlefield of the American Civil war. His phraseology, like that of many Irishmen of his date, rather tends towards polysyllables; still, the expressions that he uses show that his letter, though appearing in the *Waterford Citizen,* has not been altered or edited, but remains just as he wrote it. He tells us that he was chosen leader of the defense, and that he directed the others to let the Piedmontese into the house, and to reserve their fire until they caught them coming up the stairs. But the Piedmontese were already firing at them, and as soon as they broke down the door calling for surrender, six of the young volunteers answered back, after which the Piedmontese "ceased firing and got some more assistance." It was when this assistance arrived that their corner became really dangerous; the assailants inside the house were firing volleys up the stairs and the Irish boys were firing down on top of them, while the assailants outside were riddling the windows with bullets. The boys defended themselves desperately, but Corporal Synan was seriously wounded (apparently a second wound) and Clooney and others had been hit Still Clooney claims that he did not agree to discuss any terms of surrender until after the stairs were gone and the shutters blown in, and he feared the enemy were about to set fire to the house. It was only then that he agreed to parley, and it has always been remembered among the survivors that the Piedmontese sergeant who accepted their surrender tried hard to convey to them his appreciation of the defense that they had made.

But their most trying moment was yet to come, and the following incident shows how very little mercy the Papal rank and file expected from the Piedmontese. Once prisoners they were all taken off to one of the palaces where they found their six companions of the St. Angelo guard already awaiting them, sitting on the stairs. There seems to have been some doubt among all the volunteers, and even in the mind of a priest who was with them, as to whether they, as foreign mercenaries, would not be shot, in spite of the promises made when they surrendered; and this suspicion became a certainty when soon afterwards they received orders to go down into the street four at a time, and as each four emerged from the doorway, those still in the room above heard a volley ring out outside. This left them no doubt that their friends were being executed and that their own turn was about to come. So genuine was this belief that "they all arose and gave their money to the priest, who standing up, gave a solemn absolution to all present, including an Italian captain." In reality, however, it was an entirely false alarm due to the fact that some of the prisoners still possessed loaded weapons which they were ordered to discharge into the air as soon as they got outside. The mistake was soon rectified, and they all came down.

So much for the St. Angelo guard: about the remainder of the company we have very few details. Some portion of it remained in the fortress with the men firing down at the oncoming Bersaglieri as they dodged from door to door along the straight Corso. Lieutenant Luther is said to have shown splendid coolness in directing their fire, and two of the privates, Diamond and Lyons by name, distinguished themselves for the same quality. Diamond had been in the Royal Navy, and Lyons was an exceptionally good rifle-shot; so, being both men of some experience with fire-arms, they soon succeeded in discarding their old smooth-bore muskets in favor of two rifles lent them by Italians. With these the two comrades made so many hits—though according to one account Diamond was wounded—and showed such cool indifference to danger that their conduct was noticed, and their names favorably mentioned. As a result Diamond was awarded the Cross of St. Sylvester,[157] and eventually Lyons also received the Cross, though not until about a year later.

Towards the end of the afternoon, when it was expected that the fighting would be renewed, some of the Irish were sent out of the fortress and posted in the Palazzo Donnini so as to break the attack on the fort.

The Irish accounts—as was likely—all agree that the volunteers were furious when they heard of the , surrender; Brother Howlin, who was among them, says that they were tricked with false promises—a procedure very common on such occasions—and Clooney says that they offered to hold the fort singlehanded. However, we need not rely on Irish sources to know their state of mind; General Schmidt's report stated in very definite terms that he had been obliged to give in because his men and officers refused to fight: but he made his exceptions, and the following is the account of the causes of the surrender issued by General de La Moricière in his official report of the campaign:

> General Schmidt, in a private report to me partly attributes this result to the spirit of insubordination which became manifest during the action, in the 1st Battalion of the 2nd Foreign

---

[157] There were two men called Lyons in the battalion, James Lyons and Tom Lyons, brothers, one of whom fought at Spoleto, and the other at Perugia.

Mr. Bergin, whose father knew them both, and was a lifelong friend of James, says definitely that it was Tom, the old Bengal artilleryman and veteran of the Indian mutiny, who fought at Perugia.

Peter Diamond's name was remembered above all others by the veterans. Although of a delicate constitution, and the only support of his sister, he volunteered for this campaign, and showed the most profound earnestness in the cause. Around his name there have clustered various anecdotes, most of which cannot now be tested; according to one he made his escape after being taken prisoner; according to another version he was wounded at Perugia. Only two or three years later he died, owing, it is said, to the strain of the campaign, and so much sympathy did his death arouse that a small leaflet was printed giving an account of his life.

Regiment. The Irish company and the majority of the 2nd Line Battalion [Italian] alone showed themselves determined to do their duty.[158]

---

158   Rapport du General de La Moricière, p. 19.

CHAPTER XIII.
After the Fall of Perugia.

Perugia was now in the hands of Generals Fanti and Della Rocca and they at once proceeded to make use of it as a base for further offensive operations. De La Moricière's center of concentration was Foligno, only about twenty miles distant from them, and he had only 6,500 men to oppose to their 12,000 picked troops.

But de La Moricière was already gone. On September 11 he had heard of the invasion, and at noon had issued his orders to concentrate: he proposed to march northward to Ancona in two columns; the first about 3,000 strong under his own command; the second about 3,500 under the Marquis de Pimodan a day's march in rear.

On September 12 the concentration took place, and the troops started; on the morning of the 13th his first column had reached Serravalle, a day's march distant, and on the same evening it started for Tolentino, while the second column under Pimodan left Foligno and started for Serravalle. On September 14, the evening when Perugia was captured, de La Moricière was already on his way to Macerata, and Pimodan on the way to Tolentino; a good deal of the marching was done by night. They were thus about two and a half days in front of Fanti, but he meant to pursue them at the top of his speed.

After the Fall of Perugia

De La Moricière had hoped that Perugia would hold out for several days so as to give him a longer start.

Meanwhile his true danger was coming from the north, from Cialdini, who with 17,000 men had crossed the northern frontier at La Cattolica and was moving along the coast road of the Adriatic to cut him off from Ancona. Cialdini had started on September 11, and on the same afternoon had attacked Pesaro, the most northern town in the state. At 9 a.m. on the following morning Pesaro had fallen—though not until after a stout resistance—and on the same day Fano had surrendered without fighting. On the 13th, Cialdini occupied Senegailia, and the Pontifical troops in the northern provinces completed their retirement into Ancona. Their only approach to success had been a rearguard action at St. Angelo, where Colonel Kanzler had rather skillfully extricated his column from the pursuit of the Piedmontese cavalry. It is true that he had lost some 150 men taken prisoners, while only accounting for about half a dozen Piedmontese, but still Kanzler himself had displayed skill, and his troops, especially two or three Italian Voltigeur companies, had shown steadiness, and claimed to have repulsed three cavalry charges.

By now it was abundantly evident that there would be no serious resistance except in Ancona. All these little garrisons dotted all over the country to form centers for suppressing rebels and repulsing raids were no more than a mouthful to be swallowed at one gulp by a regular army such as that of the Piedmontese. Thus Orvieto had surrendered without the loss of a man: in Perugia 1,600 trained troops had only held out for three hours of street fighting and twenty minutes of bombardment; in Fano 250 men had surrendered without there being a casualty on either side. The only real resistance had been that of the citadel of Pesaro, where the Marquis Zappi, a gallant old Colonel, backed by a vigorous delegato, Monsignor Bella, and a garrison of 1,200 men had held this mediaeval citadel against artillery fire from 3 p.m. until dark on September 11, and from 3 a.m. until nine on the following morning. He had only three old iron guns in the place, and it is said that they were all out of action before he gave in.

These small garrisons had practically no chance of success, and they were perfectly aware of the fact; they were defending old mediaeval citadels, incapable; of standing for more than a few hours against modem artillery, in towns where most of the people were disaffected. And in spite of Cardinal de Mérode and the civil administrators, the bishops and Monsignori who ruled these provinces in the name of the Pope were nearly all in favor of a surrender, because they realized that the Papal government was no longer possible; but even in hopeless cases it was necessary to make at least a formal resistance to show that the Pope had not resigned his claim to authority, but had been driven out by the Piedmontese.

Already, at the end of only the first four days of regular war, there remained no hope of serious defense except in Ancona. If Ancona could resist, and gloriously resist for some weeks, and an appeal be made to Catholic Europe for rescue, there might then. be a hope that the sympathy aroused would compel one or other of the governments to intervene.

We now turn once again to the volunteers to inquire how they were faring during this débacle.

Many of them seem to have been in the highest spirits—a state of mind partly due to ignorance of what was actually occurring and partly to a feeling of general excitement at the idea of adventure. But O'Reilly and the senior officers were perfectly aware of the hopelessness of the situation; in fact—as the following quotations will show—there is almost a pathetic contrast between the joyousness of the men and the gravity of those who understood.

It will be remembered that O'Reilly after returning from a visit to Ancona, had been sent to Foligno with No. 4 Company, consisting of Captain Kirwan, Lieutenant Carey, and 2nd Lieutenant D'Arcy, and 130 or more men; he had been sent there on the 10th to take command of the place with full powers—a state of siege having been declared—and his men were to garrison it during the concentration; the other two companies, Nos. 2 and 3, remained at Spoleto, about seventeen miles off. De La Moricière had decided to take this Irish company with him on his march northwards, but gave orders to reduce its numbers to 100 men, presumably because he hoped that a reduced number of men might be more fully equipped. But even as it was, these hundred-and-odd Irish were obliged to march without cartridge-pouches, and in their stead were provided with a large kind of bag or haversack in which they carried both their cartridges and their rations; in fact they had sixty rounds of ammunition in the same bag with their dinner, which is an unfortunate preparation for a forced march of four sweltering days during preternaturally hot weather.[159] However, such small matters were of no importance just then, for de La Moricière's order had created tremendous competition to be among the hundred chosen. The following is the description which O'Reilly gives in his diary:—

> This day I received my orders. I was to reduce the strength of the company in Foligno to 100 men, by withdrawing any weak men, and it was to march in charge of some artillery; I was to return to Spoleto, Foligno being abandoned, take back any ammunition there, and if attacked defend it as long as possible. I begged to be allowed to march forward with the column, but the General refused, saying: "Il faut un peu de commandement partout." The delight of the Irish at being told they were to march where

---

[159] As matters turned out, they did a good deal of their marching by night.

> there was sure to be some fighting was indescribable. It was with the greatest difficulty the draft of men to return could be made out, for it was supposed still that the French would come up and that those in rear would see no fighting. Those indeed in command felt from La Moricière's silence on the subject that the belief in French intervention was gone, but all spoke as if the only fighting was to be at the front. At 8 p.m. all commanding officers attended La Moricière's "rapport," and we received our last orders. I saw the Irish company ready to march and then went to bed. I was to return to Spoleto next day.

As a matter of fact the volunteers got leave to include others, so when the company started it consisted probably of at least 105 privates and three officers, and marched off with every man cheering.

But on the previous page of his diary O'Reilly speaks of the officers, and one gets a very different picture:

> *Wednesday*, 12.—Troops arrived in Foligno every hour, and it was known the whole were marching forward on the road to Macerata. La Moricière himself, with his staff, arrived from Spoleto; little was said, but it was known the French were not advancing in support, and the whole brunt of the Piedmontese attack would have to be borne by the little Pontifical army. None, however, seemed disheartened; they were ready to fight, and thought little of consequences. Prince Odescalchi commanded the heavy cavalry. Pimodan I met on horseback in the street; he seemed sad and more than usually

taciturn. The difficulty of organizing anything like a transport or commissariat service and the sense of the little reliance to be placed in the Italian portion of the army weighed upon him. He asked me where I was going. "To try and get carts to carry ammunition." He smiled sarcastically: "And do you imagine you will get any? No, no; you will get nothing in this country in such a hurry; we can get nothing done." We shook hands and parted. This was the last I saw of Georges Pimodan, as brave a soldier as ever lived. He had labored with all his soul to organize an army, and his heart sank when he 'found it impossible; the faults of the Italian administration were too fatal, and every branch broke down when called upon to work. At the "Rapport" at night the Intendente told La Moricière it was impossible to provide bread for the troops! The general had given orders long beforehand but nothing had been done, and three thousand men were to march at midnight by a mountain road over the Apennines. The fiery energy of La Moricière, however, prevailed; he threatened that the Intendente should be imprisoned if the bread were not forthcoming, and it was promised. One Intendente who had failed to victual Perugia was put under arrest and handed over to me to be kept prisoner.

Poor Pimodan was more or less right when he doubted O'Reilly's being able to get carts for the ammunition. O'Reilly spent all the 13th searching for them, and it was only in the evening that he was able to start off his convoy and see it safely out of the disaffected town of Foligno; after which he himself cantered ahead of it and soon covered the seventeen miles back to Spoleto.

CHAPTER XIV.

An Impasse.

Spoleto is an old Umbrian walled city of about 8,000 inhabitants, built on the side of a hill so steep that the back-door of a house often has to be on the first floor. It stretches about three-quarters of the way up the slope to a small plateau upon which is the piazza or square. But from that square there rises again the summit of the hill, surrounded by an oval of massive walls varying from about twelve to twenty feet in height, in whose center stands the castle—a great, massive, oblong building, with six square towers which dominate the scene. This old castello—or "Rocca"[160] as it is more often called—dates originally from Theodoric the Goth, but the present building was erected by the Spanish Cardinal Albornoz in the fourteenth century. In it were stationed the two Irish companies, and the various other units at the depot, and in one of its square towers were situated the C.O's quarters occupied by Major O'Reilly.

The first point that strikes one about the situation in 1860 is—What a curious position for an Irish country gentleman to find himself in, at only two and a half months' notice! He was

---

160 The term "Rocca" is said to be almost always the mark of an old Etruscan citadel. Of course, the whole town of Spoleto gives evidence of its great antiquity; it was an important place in the days of ancient Rome, and also no doubt much earlier. Some of its walls contain remnants of Pelasgic work.

suddenly put to defend an oval ring of mediaeval walls with the most heterogeneous string of mixed details that has ever been seen.

The fact was that Spoleto did not contain a garrison in the ordinarily accepted meaning of the term. It was mainly used as a depot or barracks for training recruits, and a place for those men who had been left behind by their battalions because they were physically unfit for active service. Thus O'Reilly found himself in command of men belonging to seven different units and six different nationalities, with four different languages, and at least two systems of drill. According to O'Reilly's report they numbered as follows:—[161]

---

[161] O'Reilly apparently omits the officers in his calculations, which were written from memory after he had left Spoleto—I include them. General Brignone at first reported the garrison as 600 strong, but afterwards gave rather higher numbers. He includes non-combatants, for whom no doubt he required rations.

Commanding Officer (1) ...	1 officer	
Military Intendente ...	2 ,,	
Chief of Administration ...	1 ,,	
Papal Gendarmes (Italian) ...	4 ,,	94 men
Native Infantry (Italian) ...	1 ,,	15 ,,
1st Foreign Regt. (Swiss and Bavarian) ...	2 ,,	182 ,,
Foreign Carbineers (Swiss) and Bersaglieri (Austrian) ...	15 ,,	25 ,,
Battalion of St. Patrick (Irish) ...	15 ,,	330 ,,
Franco-Belgian Battalion ...		23 ,,
Artillery (Italian) ...	1 ,,	32 ,,
Engineers ...		13 ,,
Men who were ill—in hospital. or elsewhere *in the town* (not with O'Reilly) ...		51 ,,
	27 officers	765 men

He also adds in three Greek civilians, two officers' wives, an Irish civilian servant, and two others, which brings the total up to 800.

It will be seen that he includes 51 sick men who were in hospital in the town. And he gives the Irish at 346 (including O'Reilly which seems impossible for two companies (whose normal strength was only 140), together with about 30 Irishmen of No. 4 Company.

My own belief is that in Spoleto there were about 312 Irish privates and N.C.O's and about 15 Irish officers (including the chaplain and doctor). There were also some foreign non-combatant officers and men attached to the Irish for pay, clothing, drill, etc. V. Appendix.—*The number of Irish who fought in this campaign.*

Irish companies (Nos. 2 and 3 and the men weeded out of No. 4 Company), about 300 men and 15 officers. Franco-Belgians and Tirailleurs left behind by their battalions as being unfit, about 23 men.

Austrian and Swiss Carbineers also left behind, about 12 men.

Swiss recruits of the 2nd Foreign Regiment—the latest enlisted, about 160 men and 2 officers.

Italians, a few gunners, gendarmes, and engineers, about 150 men and 6 officers.

Total, about 645 men and 23 officers.

It will be seen that O'Reilly had practically no trained regular troops. And his Italians, as it turned out, were useless owing to their nationalist sentiments. But even his best units were very unequal in quality.

The Irish companies contained a certain percentage of old soldiers, Royal Irish Constabulary, militiamen, and others, who knew something of drilling and musketry, and also some men who were accustomed to a shot-gun; but very few of the others had had much training. However, they were all keen to do their best.

Of the Irish officers, however, hardly any had been gazetted to their existing rank for more than about five weeks, and the two captains for only eleven days.

The twenty-three Franco-Belgian Tirailleurs included as usual a large proportion of Vicomtes and Marquis, very smart in their blue tunics. Some were in cells for small military offences. But O'Reilly released the delinquents, added one or two Irishmen to the squad (including "young MacDermot" who apparently was a corporal), and placed the whole under the command of one of his trained N.C.O's, Sergeant Townley. Being armed with nice little Minie carbines instead of the ordinary smooth-bore musket, they proved a very valuable addition to his force, especially as they, like the Irish, were full of enthusiasm at the idea of fighting.

The few Austrians and Swiss Carbineers were also a valuable asset, because of their having rifles and because of their good spirit.

So far, then, however slight the training and however bad the weapons, there was reason to hope that these men would show fight; but from the remainder of the force it was impossible to expect as much.

The 160 Swiss and Bavarian recruits were largely composed of the very last arrivals, in fact, of boys who had hardly any training at all, and, with them there were only two of their own officers. They were, moreover, the recruits of the worst battalions in the Papal service—of those that surrendered at Perugia and ran away at Castelfidardo. But the astonishing feature of the fight at Spoleto was that these boys showed a wonderfully good will and fired away at the Piedmontese all day, although, as was afterwards discovered, they did not succeed in hitting a single man.[162] The steadiness of the raw recruits at Spoleto seems to show that it was their officers and their training which were at fault elsewhere, because they showed no signs of weakness under O'Reilly.

The last contingent on the list is the 150 Italians, mainly police. Of these, it may as well be said at once that they were practically non-existent, and it is not very astonishing. They could hardly be expected to offer their lives and to join this polyglot string of units for the pleasure of shooting down men of Italic speech, the champions of the great nation which was already in existence for any Italian who had eyes to see—a nation still inanimate it is true, but with the blood everywhere

---

[162] This is a good instance of the worthlessness of their weapons—the smoothbore muskets. They were opposed to the Bersaglieri who were firing on them from a hillside (Monte Luco) about 450 to 500 yards distant, and it is evident that the musket bullets must nearly all have fallen short while those from the Bersaglieri rifles searched the place. Although the Swiss and Bavarian recruits did not know how to shoot, if their bullets had reached the hill, they would almost certainly have hit someone, were it only by accident, during the course of the day.

An Impasse

beginning to pulse in its veins.

Thus the men under O'Reilly's command could be roughly classified under two headings, the same two which applied to the whole of the Papal army, whether at Perugia, at Castelfidardo, or at Ancona, namely, the men who were ready to fight and those who did not intend to do so. It was an exceptional situation, by no means inspiring.

The truth was that there was no chance of achieving any result of real value at Spoleto.

When de La Moricière had marched northwards he had left in Umbria four main centers of defense; Perugia, defended by about 1,650 regular troops, with orders to hold out as long as possible, so as to delay the Piedmontese; Orvieto, whose rocca was easily tenable against irregulars without artillery, such as those of Masi; Viterbo, whose citadel was garrisoned by six or seven hundred regulars in order to preserve it permanently to the Papal State—for it was within the ancient patrimony of St. Peter; and Spoleto, strategically the least important of the four, with its string of depot details. To the first three de La Moricière had issued stringent injunctions to hold out to the death; but Spoleto was not considered equally defensible either by him[163] or by Fanti. And, in any case, so far, his hopes had certainly not been realized: Perugia had surrendered after about three-and-a-half hours' fighting; Orvieto had been given up without a blow, by desire of the Papal delegate in authority there; Viterbo had not yet been approached, but its garrison showed signs of discouragement, and, as matters turned out, eventually retired from the place without firing a shot, though only attacked by irregulars without artillery; as for Spoleto it remained to be seen what O'Reilly would be able to do.

---

[163] "Do not send munitions to Spoleto... Withdraw nothing from Viterbo and tell the garrison there to defend it to the death." *De La M's telegram,* Sept. 12th. Two days later he wrote: " Tell General Schmidt to take a grip of the walls of Perugia and hold out to the last." He had already expressed similar desires as to Orvieto. *V.* A.S.R. Com.-gen. tr. pont. 3155.

Everywhere both officers and men of these garrisons were discouraged by their knowledge that successful resistance was impossible. The Piedmontese had sent into Umbria a force of 12,000 picked men; and they had, as a matter of fact, in Northern Italy, an overwhelming army nearly 200,000 strong. The Papal troops were thus fighting only for honor, and in many cases their resistance was little more than formal.

Now that Sicily and Naples had joined in the national movement it was impossible for the Papal State to remain a separate unit in the very heart of United Italy. The great work was virtually accomplished.

At Spoleto the defense was perhaps more hopeless than elsewhere because the old castle was to be attacked with artillery, and it possessed no means of retaliation. It could be battered at short range by the Piedmontese, because it had no guns with which to keep them at a distance, and very few rifles. In fact, it was almost harmless at any range outside two hundred yards. Its chances were far less than those of Orvieto or Viterbo which were attacked only by irregulars without artillery and far less than those of Perugia which possessed guns.- O'Reilly, of course, realized this and he had sent an officer to Cardinal de Mérode, asking whether there was any prospect of Spoleto's being relieved, and offering, if necessary, to hold the castle to the last and then blow it sky-high before the Piedmontese got in. In reply he received the following telegram:—

> L'Officier qui va rejoindre Mortillet m'a fait votre commission; si le telegraphe transportait les larmes, vous verriez les miennes sur ce papier. Je ne puis que vous dire, faites votre devoirLa vraie recompense n'est pas pour le plus fort; je dis faites votre devoir ni plus ni moins.

An Impasse

—Mérode.[164]

O'Reilly received this reply as he was going round the walls: "I crushed it in my hand," he says, "and said nothing. It told me we fought without hope, but it did not tell me what I wanted to know—to what extremity I was to carry resistance." It told him, in fact, that de Mérode would not relieve the place; that he would not take the responsibility of authorizing a surrender; and yet that even he would not order the garrison to sacrifice their lives.[165]

"It told me that we fought without hope." This must have meant, of course, the end of his patriotic hopes and dreams. He was face to face with failure, and probably with disgrace. He had enough arms to make resistance obligatory, but not enough to make success possible. And yet he was aware that as a result his good name would be a target for the most poisonous slanders as soon as he arrived home. In the meanwhile his young newly-wedded wife was with him in Spoleto, and this was an additional cause of anxiety.

He was to fight solely for honor—so he says in his diary— and there was a certain pathos in this sentence, coming from an Irishman of that day. No tangible advantage could be won for his side. This fact, one sees, was also realized by Monsignor Pericoli, the Papal Delegate and civil administrator of the Province, for he telegraphed to Cardinal de Mérode in the following terms:—

> To the Minister of Arms—Rome.
> The Headquarters at Foligno is withdrawing to Spoleto with the Gendarmes and other

---

[164] "The officer who is going to join Mortillet gave me your message: if the telegraph conveyed tears there would be some of mine on this. I can only say, do your duty. The true reward is not for the stronger: my reply is, do your duty, neither more nor less."—Major O'Reilly's Diary.

[165] O'Reilly was told by a Piedmontese officer that this telegram was intercepted by their agents, but that they let it through, no doubt perceiving its significance.

soldiers. Foligno is in a state of insurrection at the advance of the Piedmontese troops. The garrison of Spoleto and I have decided to occupy the Rocca [citadel] in order to maintain the dignity of the Government. Spoleto and the Province are quiet.

L. Pericoli, *Apostolic Delegate*.[166]

The utmost hope, then of Monsignor Pericoli, was to maintain the dignity of his government, so as to prove that it gave in only to force. And this fact, that the fight was only for honor, is in reality more clearly perceived by the late Colonel Vigevano—himself an officer and an Italian gentleman—than by any of our own writers. He says: "With them [the troops from Foligno] was the Irish Major O'Reilly, who had orders to complete the work necessary for putting the Rocca of that town [Spoleto] in a state of defense, and to make a resistance that would safeguard their military honor."[167]

And when describing the garrison of Spoleto, he says:

This conglomeration of soldiers of various nationalities, of dissimilar systems of instruction, and belonging to different corps, could not possibly have a high coefficient of resistance, and Major O'Reilly had no illusions upon that score; but he was determined not to give in unless with honor, his only regret being that he had not with him the other six companies of his battalion of St. Patrick; these, however, had been ordered away for separate service, one at Perugia, one with Cropt's column, and the other four in Ancona.[168]

---

166   A.S.R. M.A. Aff. Ris. 1964.
167   M.G. Uff. Stor. p. 238. *La campagna delle Marche e dell' Umbria*. By Col. Vigevano.
168   M.G. Uff. Stor., p. 255. *Ib.*

CHAPTER XV.
Last Preparations.

O'REILLY'S feverish work during the next two days is best described in his own vivid sentences:

> 14th.—Early the next morning I set to work to see after provisions, etc., for the citadel. I was, of course, senior in command of the subdivision of Spoleto, as it was called, all other troops having marched away. La Moricière had told me he had given orders to have the Rocca provisioned, but I found nothing had been done. However, the Intendente Viviana[169] said he would furnish the provisions of which most were in store in the town, on my order backed by that of the Delegate or Governor. This was readily given, for the Delegate was a man of energy, and, after a great deal of trouble getting carts on requisition from the municipality, biscuit, coffee, bacon, some wine, and a little brandy [were] carried up; but it was with great difficulty that I got any torches for night, and all attempts to have the pipes

---

[169] The Intendente's name was Viviani. Throughout all this passage will be found small mistakes, showing probably that Major O'Reilly had not revised it before he died.

which filled the well in the Tower with water from the aqueduct, failed. For my own provisions I had a couple of dozen cases of preserved meats which I had sent up to the two rooms in one of the angle towers which I occupied, and a couple of flasks of brandy.

The delegate told me in case of attack he would take up his quarters in the citadel as the civil representative of the government of His Holiness, and would leave all the military authority to me. He was a priest and showed great coolness and courage, but, unfortunately, from insisting on keeping guards at all the gates of the town and remaining himself in the palace until the last moment, he obliged me greatly to exhaust my best soldiers as double guards at all the posts under experienced non-commissioned officers—and, in momentary expectation of an attack or a rising, I could send none others—kept nearly all my old soldiers on constant duty for three days before the siege began.[170]

In the afternoon of Friday, 14, I received a message from the Delegate that he wished to see me. He told me that he had received private information that Perugia had fallen, and the enemy were at Foligno or near it and might be expected soon; but not to publish this news.

15th.—Saturday.—...On Saturday evening the Delegate came and took up his quarters in the citadel, and all the guards were withdrawn. Mobs assembled in the streets and some of the guard were threatened with attack; but their

---

[170] This was evidently to show that the Papal government had not resigned a shred of its authority until actually forced to do so by the Piedmontese.

determined bearing checked any such attempts. As the main guard was retiring from the palace of the Delegate across the square, two Piedmontese lancers galloped into the square; the sergeant of the guard and a soldier instantly covered them and were about to fire but the officer seeing that the two lancers were alone and had only galloped up in bravado, struck up their muskets, and told the men not to fire, and marched his guard quietly to the citadel; with great difficulty I got all the ammunition carried from the guardroom outside the tower in which it was into the large lower room, vaulted in the central tower, which had been prepared as a magazine.

The night was spent by the Irishmen chiefly in dancing and singing; no authority could get them to go to bed, they were so excited at the prospect of fighting.

Dancing and singing are a common enough item in the history of both Irish and British regiments before a battle, but in this instance it produced a rather curious result. Some fifty years later, when talking with Signor Santorelli, a Spoletan who as a boy had been one of the excited spectators of the engagement and had afterwards published a small book on it, he told me that on the night before the battle the Irish had held an orgy and that it was said that they had fallen out amongst themselves and had had about a hundred killed and wounded; he asked me whether this was true; he had mentioned it in his book. That Santorelli, who was quite an educated and an impartial man, should have credited such a story for a moment merely shows the kind of atmosphere which had been created around the foreign troops of the Papacy. Almost anything could be attributed to them. A few months later, when talking to Edward Dunne, a Dublin veteran, I asked him what had occurred, and he smiled at the reminiscence. "Well, we did our best to keep up pleasant," he said, and added that the evening had gone by without any kind of untoward event whatsoever, and with everyone trying to contribute to the general hilarity, as is the usual object on these occasions.

In the meantime, while O'Reilly was toiling, perforce in the volunteer manner—that is to say, obliged to make enthusiasm take the place of previous preparation, and to overwork his few trained men—the Piedmontese were continuing their advance with the systematized care that characterizes a well-organized army, and perhaps especially General Fanti. In this campaign he had an easy task to perform, but certainly it was most uncommonly well done; every possible contingency had been thought out beforehand.

On September 14th, the same evening on which his main body captured Perugia, his reserve division had already advanced as far as Colle Strada on the way to Foligno. On the 15th he advanced to Foligno, which he now made his headquarters for future operations. His intention was to wheel to the left and follow in de La Moricière's tracks northwards. He would thus be fairly close on the rear of the small Papal force of 6,500 men with his own 12,000 men; while Cialdini was straining every nerve to get across their front with his 17,000 men; and, finally, Cadorna with his central division of 7,000 men was moving down the mountain valleys towards their left flank; on their fourth side was the sea. De La Moricière was to be surrounded like a fox that is chopped in cover, unless he could get through to Ancona.

Meanwhile there remained one small Papal force which had to be polished off while Fanti was going northward, namely O'Reilly's 650 men in Spoleto. In itself this force was unimportant, but Fanti thought that it might perhaps become a center of reaction, especially if it were able to act in concert with Rome or Ascoli, for the mountaineers of Ascoli were devoted to their kindly old Pope.

On September 15 Fanti told off the column which was to take Spoleto and leave a garrison there. It was to consist of the following troops under command of General Brignone:

The 3rd Regiment of Grenadiers (4 battalions) 1,576 men; the 6th Field Battery of the 8th Regiment (6 guns), 173 men; the 9th Battalion of Bersaglieri, 419 men; 2 Squadrons of Niçois Cavalry, 222 men. Including staff this gives a total about of 2,400 men,[171] which was evidently considered by Fanti enough to make the matter certain.

On September 16 (Sunday), at 2 p.m., this force started from Foligno on its road for Spoleto. It was, of course, an excellent little mixed unit composed of some of the best corps in the Piedmontese service: at its head moved a squadron of the blue Niçois Lancers; then came three battalions of Grenadiers; then the battery of 12-pounders ("da 16"); after that another battalion of Grenadiers; then the 9th Bersaglieri,[172] full of enthusiasm to rival the glory won at Perugia by their companions in the 16th Bersaglieri; and lastly the other squadron of Niçois Lancers.[173]

Their advance continued without incident until the column reached the village of S. Giacomo, about six kilometers from Spoleto; at that point a patrol was sent out under Lieutenant Bouvier to reconnoiter the town, and it had not gone far before it fell in with a patrol of mounted Papal gendarmes.

---

171   The above are the figures given in the Italian official accounts. They have been very much disputed, especially those relating to the artillery. O'Reilly was under the impression that he had to do with the whole of Fanti's division, which of course was a perfectly possible supposition, though mistaken.

172   The word Bersaglieri means marksmen or sharpshooters, and the name rather corresponds to that of Tyrolese Jägers. These battalions were composed of picked men who were put through a special course of training, often moving at the double.

173   M.G. Uff. Stor., p. 243.

Plan of the Castle of Spoleto

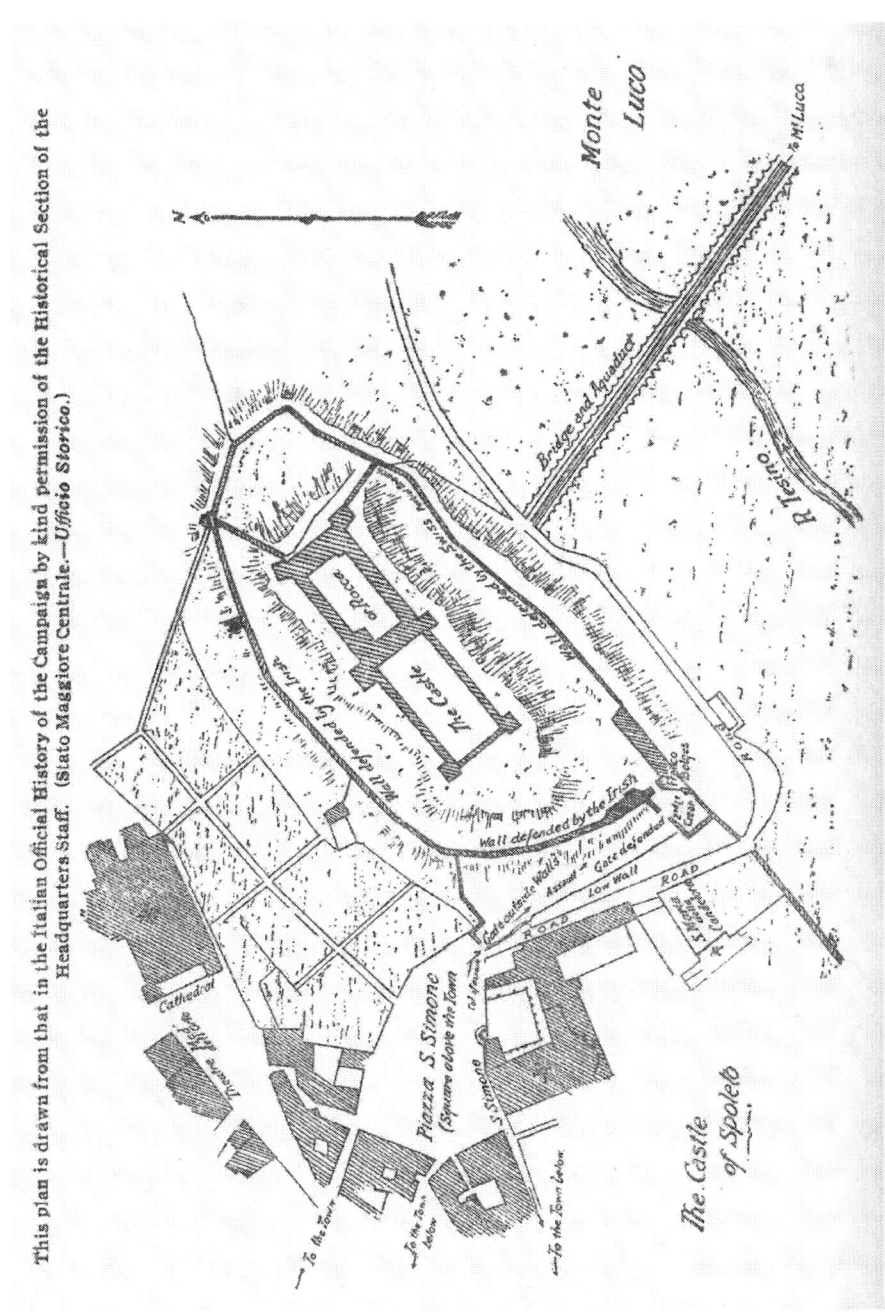

(Scan curtesy of Ave Maria Canizaro Library)

Bouvier's lancers at once attacked them, and chased them back to Spoleto, taking five prisoners, and he was able to report to Brignone that the town was now free of the Papal troops, which had all retired into the citadel.[174]

> I have sent to-day a column against Spoleto to take that castle with the 400 or 500 men who are garrisoning it. I hope the coup will be accomplished to-morrow morning.—*Extract from General Fanti's Letter of September 16 to General Cadorna.*[175]
>
> I have sent Brignone with a Regiment of Grenadiers (3rd), a battalion of Bersaglieri (9th), two squadrons and four pieces of artillery to Spoleto to make the Rocca capitulate; in it there are 600 Irish, Germans, etc., and a small squadron of dragoons.[176] I think it will be a matter of a couple of hours.—*General Fanti's letter of September 17 to General Cadorna, O.C. 13th Division (Active) on the road to Cancelli.*[177]

From these quotations it will be seen that Fanti regarded the Rocca of Spoleto as good for only about two hours; and it seems only fair to quote these extracts in view of the writings of various anonymous journalists who, after the war was over, levelled every sort of slander against the defenders in order to

---

174   It was evidently two of these lancers who galloped into the square as described by O'Reilly; but he put the episode on Saturday 15th, whereas it occurred on Sunday 16th—doubtless a slip. According to one account the townsmen witnessed the chase of this mounted gendarme, who arrived galloping for his life with the lancers on his heels, but he succeeded in escaping them when he got through the gate by leaving his spent horse at the foot of the hill, and making his way up the steep short cuts on foot.
175   Uff. Stor. I. p. 209.
176   They were not dragoons but some mounted gendarmes, Italian policemen who, as it turned out, refused to fight against their fellow-countrymen.
177   Uff. Stor. I. p. 210.

Last Preparations

ruin their lives.

During these last days O'Reilly on his part was straining every nerve to make the most of his very insufficient resources. He had an old castle and its surrounding courtyard wall to defend; of effective weapons he had very few; about sixty rifles or rifled carbines, the rest being smooth-bore muskets which were useful only at short range; he had no artillery that could be called such.[178]

The old castle stands square and massive on its hill overlooking the town, and its walls are very thick; in fact, it was strong enough to stand for some time against the Piedmontese artillery of that day. Nevertheless it was undefendable, because it had, as O'Reilly says, "few windows, and those high up, and no parapet or means of standing on the roof, no loopholes and no flanking fire whatever; consequently once in there we might be blown up at leisure by the enemy whom we could not touch from the inside." And his judgment is not questioned in the Italian official history of the campaign.

Around the castle, however, was its oval ring of courtyard walls, varying in height from about twelve to twenty-five feet, or even higher, and built at a lower level of the hill than the castle itself—about thirty-one feet lower according to O'Reilly. These walls were sound; in fact they were the best asset on the side of the defense. They were strong enough to stand artillery fire, at all events for a day or so, and they were well loop-holed and gave good protection.

O'Reilly's difficulty was that he had so few weapons that would be effective at more than 150 or 200 yards' range. He had no guns with which to reply to the enemy's artillery and could only do so by rifle fire; and he had hardly any rifles. The twenty-three Franco-Belgians were armed with Minie rifled carbines, and the twelve Swiss Carbineers and Austrians had rifles, but beyond these there was no one trained to their use.

---
178   Maj. O'Reilly's diary.

One day, however, shortly before the war began, when going round the armory with General de La Moricière, O'Reilly had noticed twenty-six rifles left there by the Italian Cacciatori regiment, and had got leave to use them; so he now served them out to the twenty-six best shots among the Irish, with orders to move from point to point and to use them "wherever they thought their fire could be most effective." These few weapons, for twelve hours, were his only possible means of replying to the Piedmontese field guns.

There were, inside the Rocca, three pieces of artillery; two six pounders,[179] and one old gun which had been kept for firing blank cartridges on festival occasions; at the beginning of the battle, an attempt was made to use the latter, but it failed.[180] Of the two six pounders one was unserviceable[181] but the other was in good order. It was not possible, however, to pit this small gun alone against the Piedmontese battery, so O'Reilly decided to place it in an embrasure beside the gateway, to repel assaults—in fact to use it rather in the same manner that a very large duck gun or blunderbuss might be used at short range. But on the day of the battle he was able to get only three shots out of his artillery, all told, and one of those three shots fell short, although he had an excellent artillery officer, Captain de Baye of the French army, and an equally good Italian sergeant.[182]

O'Reilly's task then was to keep the enemy out of that circle of walls; if once they penetrated it at any point he was lost. His dispositions were as follows.

---

179   The Italian account calls them eight pounders; but that is about the same weight as an English six pounder; the old Piedmontese libbra was about 1/3 of a kilogram, or 11/13 of a pound.
180   Conversations with Signor Santorelli.
181   M.G. Uff. Stor., p. 255 note. Colonel Vigevano's description
182   The Italian gunners were not all good, but this does not seem to have made any difference. The officer was Baron Christian de Baye, and he had been for seven years in the French army; he seems to have been very practical, and to have stood well by O'Reilly.

The northern side needed no defense; below the wall is a bare precipice.

The western side is rather similar, but some portions of the precipice are covered with fir trees and scrub which would afford cover to an enemy scrambling up it; so O'Reilly stationed one section of the Irish along this portion of the wall.

On the eastern side the ground plunges down into a ravine at the bottom of which was a large stream bridged by a gigantic viaduct; and on the opposite side rises the wooded mountain called Monte Luco. On Monte Luco the Piedmontese riflemen could get perfect cover at a distance of only 450 to 500 yards, and could overlook the whole place while remaining invisible themselves. On this side, too, there existed two partial breaches in the top of the wall, which was still under repair, and O'Reilly believed that an escalade was possible, if not by day at all events by night.[183] To provide for this danger he had the larger of the two breaches filled with "a hastily contrived defense of bales and blankets" and told off a section of Irish to hold it; and the smaller of the breaches he allotted to Sergeant Schafter with the twelve Austrians, although this meant immobilizing twelve out of his small number of rifles. The rest of the wall on that side he lined with his Swiss recruits.

The Italian contingent was kept in reserve within the castle itself.

But it was on the southern and south-western wall that lay his real point of danger, because on that side lies the town, built on the face of the hill just below and clustering as if for protection at the foot of the castle. A hostile force in the houses across the piazza could make fairly good practice at the defenders on the castle walls, which are only about 100 to 150 yards distant. This side of the defense is the most important to note, because it is the most vulnerable and it is here that

---

183   V. Letter of Gen. Zappi. Aug. 3rd. to De La M.—A.S.R. Com. Gen. tr. pont. 3151.

Slope up to the gate attacked ↓

The Rocca of Spoleto taken from the south-east; the town would lie on the left of this photograph. It was from the town side that the assault was launched; but the most harassing feature of the attack was the rifle-fire from the higher levels of Monte Luco.

← 500 yds. range
← 600 yds. range
← Monte Luco

(Scan curtesy of Ave Maria Canizaro Library)

Last Preparations

the really serious part of the fighting took place. At the top of the town lies the piazza (square); from the piazza a road leads up about fifty yards to the nearest outstanding angle of the courtyard wall, and inclining to the right of the angle coasts along beside the wall, up the hill for another 120 yards or so, until it passes under the gateway of the fortress, which juts out across it at a right angle from the wall. It was this spot that the Piedmontese selected for their assault: to batter down the old wooden gate with their artillery and then to storm the gateway was the project on which they eventually decided when they found that O'Reilly was not disposed to give in. It was, of course, a reckless plan; to charge up that slope for 120 yards under fire all the way from the loopholed wall on their left and from over the gate in front would mean certain death for many of those who went first: but they were in the mood to take chances.

O'Reilly had foreseen that this was the point of danger, so he posted his Irish all along the wall on this side; a section just inside the gate under Lieutenant Crean, described in those days as "a stout boy from Tipperary"; and several other Irishmen on top of the wall above it, although they were exposed to fire from nearly all four points of the compass at once—from their front, from their left, and to dropping bullets from Monte Luco. At the back of the gate there is a small house with an open gallery overlooking the roadway up to it; in that upper gallery he posted his twenty-three Franco-Belgians and the two or three Irish who were with them, so that they could fire over the top of the archway at any force approaching it. In the embrasure of the gate itself Captain de Baye placed his little six-pounder to sweep the road up to it.

Just inside lies a courtyard and then a second gateway, across which O'Reilly kept two old wagons "ready to be upset to form a barricade" in case of the first being forced; and beside them he had two sections of Irish in reserve in case of an assaulting column getting through. Thus every possible precaution had been taken; but the gateway still remained the point of danger because an assaulting column would not be visible to the defenders until it came within about a hundred and fifty yards of them.

Meanwhile the Piedmontese force was halted at the village of S. Giacomo, about three-and-a-half miles off, and the townspeople flocked out to greet them, the men marching side by side with the Bersaglieri and the women strewing flowers in their way. At least, the scene is thus described by Santorelli, but one of their officers has recorded it in his memoirs that they were received with less enthusiasm in Spoleto than in any other Umbrian town. Both accounts are probably true; no doubt a crowd of rejoicing people went out to welcome the Bersaglieri; no doubt also there were many others who were not enthusiastic at their approach; for Spoleto must have perceived that under the new regime Perugia would take her place as chief town of Umbria. Perugia represented the young progressive spirit; Spoleto the old historic tradition; and a rivalry existed between them which has continued to this day.[184]

When darkness fell Brignone sent his various units to their posts; before midnight he had dispatched his two squadrons of Nicois Lancers along the Terni road, so as to intercept the garrison in case of their abandoning the castle.

On the northern and western side of the castle he had posted as reserve in the Borgo S. Gregorio, the 3rd and 4th

---

[184]   Corsi, the well-known military historian, served with Fanti's force throughout this campaign. He tells us that when the Piedmontese force first crossed the frontier and took Citta di Castello, they were received with anything but enthusiasm by the inhabitants; but when they came near Foligno they were welcomed with every sign of joy. Spoleto was never very revolutionary; Carlo Bruschi, the well known rebel leader has left us a list of all the towns of Umbria, giving the number of armed men that each could be relied on to provide, and the Spoleto figure is nil. But Bruschi was a Perugian, so perhaps he took a contemptuous view of the Spoletans, which they may have returned.

Battalions of Grenadiers and a section of artillery.

On the eastern side he ordered two companies of Bersaglieri to occupy Monte Luco, the wooded height which overlooks the ring of walls; from thence it was possible for them to see right into the place, and search it with their rifle fire.

On the opposite south-western side, where the gate is, he posted the 1st and 2nd Battalions of the 3rd Grenadiers and the other two companies of Bersaglieri in the upper houses of the town or at any points that commanded the outlets from the Rocca.

His artillery, he tells us, was thus divided: one section of two guns was at the entrance to the Borgo S. Gregorio; the other two sections were sent scrambling up the bad roads to occupy the hill of Colle Risano.[185]

All these dispositions were carried out after .midnight in the most profound silence. The defenders of the Rocca could hear movements in the darkness and occasionally a bugle-call or the neighing of a horse and they knew that their enemies were taking post all round them. But they could not tell where; in fact they never, even during the fight next day, were able to locate them all for certain.

Although no shot had yet been fired, these night hours might well be counted as the first period of the defense; to some, at all events, of the volunteers they . seemed more trying than the fighting on the following day.

---

185   Also written Colle Risciano and Collerisano.

CHAPTER XVI.

THE FIGHT AT SPOLETO; SEPTEMBER 17th.[186]

"About four in the morning," says Major O'Reilly in his diary, "the assembly sounded, and I jumped up. 1 found that, hearing the enemy moving near us, some of the young officers had hastily ordered the assembly to be sounded. I was annoyed, as it would only fatigue the men; but it was too late to stop it, and useless to try and send them to bed again, so I let them go to their posts, desiring them to lie down on the ground, and I went out myself and lay under a cloak."

At 6 a.m. a Piedmontese officer with an escort and flag of truce came up to the gate, and O'Reilly went out to meet

---

186   This chapter was read through and approved by Mr. M. T. Crean and by Edward Dunne, both of whom took part in the defense—Mr. Crean as the Second Lieutenant at the gate where the close fighting took place, and Dunne as a private, on the wall near it, at the south side. The main points were also submitted to Santorelli, who was a spectator, and to Vallini, who was an ex-Papal gunner, but too old to remember much.

The chief written authorities are : On the Papal side—Major O'Reilly's Diary and reports; Mr. M. T. Crean in the *Seven Hills Magazine* for March, 1908; General de La Moricière's report, and the short descriptions given by one or two Franco Belgians, such as that of Emmanuel de Gouttepagnon. But by far the most valuable sources were my long conversations with Mr. Crean and Edward Dunne.

On the Piedmontese side there were: General Brignone's report (repeated of course in Fanti's narrative of the campaign); the Historical Military Diary in the Italian War Office (Uff. Stor. Ip. 11); the description of the assault issued by the Italian War Office some twenty years ago, in answer to an enquiry from the inhabitants of Spoleto (marked No. 93 di protocollo, but I have not got the year); Santorelli. *La presa di Spoleto;* and Vigevano's official history, *La campagna delle Marche e dell' Umbria.*

The Fight at Spoleto; September 17th

him. All the parleying was done outside the gate in order "to prevent them seeing how weak our position was." He stated that he was the chief of General Brignone's staff; that the general was commanding a division of the Piedmontese army and had orders to require the surrender of the place. "I, of course, replied in writing that, holding it for His Holiness Pius IX, I could not surrender it to any other authority; but at his request I sent Captain Coppinger to the general. He returned saying that the general offered most reasonable terms of capitulation, but that he of course replied that surrender was out of the question ... He [the Piedmontese staff officer] urged the uselessness of resisting; that General Brignone commanded a whole division[187] with plenty of artillery; that no succor could come to us, and that I could not be expected to defend an untenable post I simply answered that he must know as a soldier I was bound to defend it as long as it was tenable, and that he would find it could be held."

Out of this interview, however, there arose one good result. General Brignone, in his usual courteous spirit, agreed to give a safe-conduct to the women who were in the castle, and consequently Mrs. O'Reilly was allowed to pass out in safety to the town. She was very much distressed at going, especially when it came to saying good-bye to her husband and the volunteers—and, in fact, turned back again after passing through the gate—but it must have been a great relief to Major O'Reilly when he could once feel that she was in safety, and that she had with her Mrs. Boschan the wife of his Austrian officer, her own maid, and an Irish civilian servant.

When the Bersagliere officer with his white flag had escorted the ladies away, Major O'Reilly went and sat down "on a knoll under the commenced bastion where the one gun was, with

---

187   O'Reilly believed, naturally enough, that he and his recruits were face to face with the whole of the division which had taken Perugia. The Piedmontese approach had been carried out in the dark.

Sergeant-Major Mulhall and one or two others. Captain de Baye urged that politeness obliged us to reply to the fire of the Piedmontese with our one gun, at least at first, although of course from its caliber, 8 lbs., it could not reach them, nor from its exposed position could it continue to be served.

> A few minutes after eight o'clock[188] the first round shot from the Piedmontese guns hurtled over my head, struck the south wall of the old tower, which it hardly marked, and rolled down to us ... De Baye ... took the match and fired the gun himself; it gave a most faint and hissing report as if the cartridge was bad and, as I afterwards heard from General Brignone, the shot fell dead a short way over the town; that was the last cannon shot. In a very few minutes the fire became warm and general.[189]

The action had thus begun very quickly; before they were well aware of it the round shot was thudding against their old walls and the shells dropping here and there, while the rifle bullets threw up pebbles and dust all around them. It seemed that the artillery fire was coming from two or three different points at once. Through his glasses O'Reilly could distinguish the guns on the Colle Risano, about 1,200 yards distant, and soon perceived that they were directing their fire at the old keep; while to the west, only about 600 yards distant, he saw another section which was aiming its shot and shells at the former powder magazine, of whose position the gunners had evidently been informed; but they made bad shooting and did not succeed in hitting it. Meanwhile, from their loopholes in

---

188 The Italian official account says 10 o'clock, but that is a mistake. O'Reilly's time is confirmed by Mr. Crean and other Irishmen there, and also by Signor Santorelli, an Italian eye witness. The *Nazione* account said that the fight began at 6 o'clock, but this is also a mistake. V. *Nazione*, Sept. 26, 1860.
189 Maj. O'Reilly's diary.

The Fight at Spoleto; September 17th

The Rocca of Spoleto; gate at which the close fighting took place.

(Scan curtesy of Ave Maria Canizaro Library)

the walls, the Irish and their companions were beginning to reply whenever they got a chance at any of the assailants who showed themselves within short range.

To the distant artillery O'Reilly could make no response at all, because he had not got a weapon of any sort that would carry 1,200 yards; and even at the other section of guns on the town side his chances were very little better, because in the year 1860 it was not very easy to hit anything so far off as 600 yards even with a rifle. Still he kept his twenty-six Irish and his twenty-three Franco-Belgians moving round from point to point on the look-out for any opportunity of scoring, but more especially to check any attempt, by the artillery, at coming closer: and it was not long before such an attempt occurred. It was made by the Piedmontese gunners on the town side, who had probably begun to realize that they were not producing any great results, and also that there was no artillery against them. The episode is thus described by O'Reilly in his diary:—

> Finding after some time that they were too distant they moved one gun a good deal nearer. Second Lieutenant Cronin, a very good rifle-shot, who was on the wall engaged with the Piedmontese sharpshooters in the woods below him, was watching the gun, when a Franco-Belge, an old African soldier, came to him and, making signs (Cronin spoke no French) that he thought he had the distance right, 600 yards, fired just as the Piedmontese artilleryman was about to fire the gun. The artilleryman rolled over and was carried into a neighboring house. Another gunner then advanced to fire the gun, but the Frenchman had in the meantime loaded. He again fired, and the second artilleryman fell. This happened a third time and the

The Fight at Spoleto; September 17th 161

Piedmontese then gave up working the gun.[190]

But far more dangerous than the artillery fire was that of the Piedmontese rifles: "they searched the whole place," says O'Reilly, and this is not surprising seeing that they were firing into it from both sides at once: from the town side they could fusillade the Irish on the banquettes at a distance of only about 150 yards, while from Monte Luco, on the opposite side, there came the dropping bullets of the Bersaglieri, to hit the defenders in the back. On the wall opposite Monte Luco O'Reilly had placed his Swiss and Bavarian recruits, and they kept up a constant fire all day across the ravine at the Bersaglieri concealed on the mountain, but—according to the Piedmontese returns—without hitting even one single man among their enemies; which is an eloquent tribute to the badness of their muskets, because it suggests very strongly that the bullets must all have been falling short.[191]

In this manner the engagement continued without any salient event on either side; but at the end of the first two and a half hours only two or three of the Irish had been wounded, and throughout the ranks in general of these half-trained volunteers their successful resistance was greatly increasing their confidence both in themselves and in their officers. They had seen some of the enemy fall—Major O'Reilly says that the

---

190 This story, resting on the authority of Lieut. Cronin, would seem conclusive; but the Piedmontese have never admitted to the loss of any gunners. Cronin may have been wrong; one knows, of course, that mistakes are easily made at a distance of 600 yards, and occur in every battle: on the other hand the Piedmontese make no mention of moving their gun forward, though this would have been a very probable step on their part.
  The above, however, is an instance of the only kind of defense that the Irish could make throughout the day—namely, using their rifles to keep the Piedmontese guns from coming near enough to batter down their walls.
191 O'Reilly attributes this to the fact that they did not venture to raise their heads to aim; but, though this may have been true at first, it was not the case later on. Mr. Crean spoke very well of them, and among other things said that he had noticed them raising a hat on a stick at one point in order to draw the enemy's fire, and then aiming at the puff of smoke. Edward Dunne also said to me: "Certainly they fought well."

first man killed was a civilian insurgent sniped in the piazza by Sergeant Lloyd—and felt that so far they were more than holding their own.

O'Reilly seems[192] to have set them a fine example throughout the day: he strolled from point to point almost regardless of the bullets that followed him about. And yet there must have been plenty of them, for when Mr. Crean was wounded that afternoon, it: was a long time before he could get to the doctor—so he told me—although he was in great pain.

At eleven o'clock a bugle sounded on the Piedmontese side, and a party under a flag of truce advanced towards the gate. This proved to be the Archbishop, Monsignor Arnaldi, who had applied to. General Brignone for a truce in order to see whether a surrender could not be arranged so as to prevent: further bloodshed. He was a venerable, kindly, old man in white robes, full of benevolent intentions, so Major O'Reilly went down to meet him at the gate, and by his wish sent for Monsignor Pericoli, the Papal delegate or civil administrator of the province, for a consultation. At this little conference of three held in the guard-house, "the good Archbishop," says. Major O'Reilly, "urged that resistance was useless; that the Holy Father himself would not wish the blood of his defenders to be shed in a hopeless. contest; that the Piedmontese general offered honorable terms and that he himself was bound to do all he could to save the effusion of blood. The Delegate behaved very well: he said he left the decision entirely to me; he was not competent to tell a soldier his duty, and as the civil representative of the government he would share our fate whatever it were." Monsignor Pericoli certainly showed a fine courage and devotion to duty in electing to remain within the Rocca

---

[192] The veterans of course spoke extremely well of O'Reilly. That was to be expected. But an Italian account, apparently written by one of his Papal officers, contains the following expression: "Major O'Reilly, though hard, was a brave man." This is high testimony, for O'Reilly was greatly incensed with. the Italian officers that day; it was inevitable.

during a bombardment.

O'Reilly replied that he could not surrender the place so long as he had a reasonable prospect of defending it; so the old Archbishop Amaldi gave them his blessing and returned to the town.

During the respite the men had got some food and were all fresh again.

At mid-day[193]—or perhaps a little earlier—the firing was resumed; and the attacking force adopted new tactics and showed increased vigor.

The Piedmontese now concentrated their efforts on battering down the gate by artillery fire. This, one might naturally suppose, would have been merely a question of a few straight shots, but as a matter of fact it proved very difficult: their gunners took some time in getting the right elevation and then, although they hit it three times, it was so rotten that the balls went clean through, making round holes without shaking it seriously. One of these balls struck a Papal soldier named Alfred[194] Chambers, one of the two or three Englishmen in the garrison, and (in the words of Edward Dunne, a veteran living in Dublin in 1921) it "took the whole side out of him." On this occasion Lieutenant Crean had a narrow escape; he had just observed Chambers and another man sitting with their backs to a wall, behind the gate, so he ordered them to move; but as he spoke Chambers fell dead, and the ball, having thudded against the wall behind him, rolled down to their feet.

It was during this period, I think, that most of the defenders' casualties occurred; but there were very few men hit—although they seem to have been reckless enough, especially when they wanted water—because on the town side, their ring of walls

---

193   O'Reilly says that the firing began at a few minutes after 8 a.m., and continued until 11 a.m., after which there was an hour's truce. Santorelli, an eye-witness, also says it began at 8 a.m., but puts the truce at 10.30 to 11.30.

194   His name is elsewhere given as John. V. *The Standard* (Dublin), 23rd February, 1929.

gave them good protection, and from Monte Luco the range was perhaps a little too long for accurate shooting in the year 1860. Still there were various casualties. In one case a Kerry man named Malachy Sheehan, by trade a draper at Killarney, was shot through the throat while firing from his embrasure, but the man next him bound up his neck with a handkerchief, and he ultimately returned to Killarney perfectly cured. This was a good piece of work on the part of Dr. O'Flynn, their military surgeon, who afterwards became a well-known figure among the Papal Zouaves. There were several others wounded during the course of the morning, but only one Irishman was killed, a man named Fleming, also of Killarney,[195] who was shot through the head. According to the story commonly told, his brother was next him, but as soon as he saw that help was useless he continued firing at the Piedmontese as coolly as before.

By about three o'clock in the afternoon, General Brignone decided that after six hours' firing it was time to try an assault on the gate.[196] He was led to this decision, he tells us, by perceiving that owing to the height at which the castle stood his artillery fire could not alone be sufficiently effective to compel its surrender. A storming-party was therefore assembled close to the piazza above the town, only about 150 yards from the gate itself. Fortunately, however, for the garrison, this movement was observed from an embrasure by Sergeant Lloyd, who at once went to warn Major O'Reilly and to carry out his dispositions. Captain de Baye was sent to take command of the six-pounder in the embrasure beside the gate. O'Reilly himself took up his station near the inner gateway so that he could watch the defenders in front, and at the same time keep his eye on two sections of the Irish whom he had posted in

---

195   No. 3 Company was sometimes called "the Kerry company," owing to its containing about thirty Kerrymen, some of whom were often in the public eye!
196   Brignone's report. Uff. stor. 21. P. 189.

The Fight at Spoleto; September 17th

reserve 'close by, as he fully expected that the first entrance might be forced.

As this assault at Spoleto was considered by the Piedmontese to be one of their most gallant performances during this short campaign,[197] it may be as well to repeat our former description of the ground: we have already explained that when a storming party left the town it would have to run up the road which passes under an old gateway outside the fortress; and then—inclining to the right and coasting along with the castle wall on their left—the stormers would have to dash up a slope for about 120 yards until their road was barred by the true gateway of the fortress which juts out at right angles across it. These 120 yards would be commanded by the fire of the Irish posted on top of the old arched gateway and along the wall at the side, by the small gun beside the gateway, and by the twenty-three Franco-Belgians under Sergeant Townley in their gallery which stands high up in a house behind it. Thus these last 120 yards would be exceedingly difficult to cross; but if the passage were once forced, its defenders would be outnumbered and lost.

Consequently, when General Brignone gave the command for an assault, headed by a storming party with axes, it was felt to be a somewhat daring attempt. Two companies of Bersaglieri (the 33rd and 35th) were named for the work, supported by the first battalion of Grenadiers, from whom the storming party were selected. Here it will be best to quote directly from the Italian narrative.[198]

On receiving the order, Colonel Boumod, who was in command of the 3rd Grenadiers of Lombardy (now the 73rd

---

[197] The survivors of the storming party and the families of the killed were all given a medal; and besides these there were eleven decorations given to officers who took part in the assault.
[198] Narrative given in the *Avenire* for September 19th, 1889.. This extract from the *Avenire* (a Terni and Umbrian newspaper) is preserved in the Archives of Spoleto it is also quoted in other newspapers, such as the *Messagero*, Nov. 13th, 1909.

battalion of the line) called his officers round him and said: "Gentlemen, this is going to be a bold stroke. The storming-party has been selected from among the pluckiest men in the regiment, But now we must have an officer to lead them."

Mozzoni, who was then only Furriere-Maggiore, at once stepped forward and said: "I should like to lead them, sir, if you would allow me the honor of doing so."

Bournod, whose object it had been to get a volunteer to lead the party, received Mozzoni's request with a "Bravo!" of approval.

The attack was at once initiated. One company of Bersaglieri took up a position opposite the gate, occupying among other points the topmost houses of the town in order to protect the assault by firing from the widows. Rifles and artillery crackled and crashed with redoubled vigor in a supreme effort to dominate the defenders. Meanwhile the column of attack was formed up under cover: in front, the storming party; next them, the 35th company of Bersaglieri, and, in support the 1st Battalion of Grenadiers. This column must have been almost as numerous as the whole strength of the defenders. Once inside the gate the castle was theirs. It remained to be seen whether they could force their way through.

Presently they started at the double: first came the storming party headed by Mozzoni waving his axe; then the company of Bersaglieri under Captain Prevignano, and lastly the battalion of Grenadiers bringing up the rear under Major Scaletta. They advanced at full speed "with admirable valor," says General Brignone, who was sitting watching them on his horse below, and this eulogy is well confirmed by the accounts of the defenders,[199] for their courage during the attack filled the Irish with admiration. They passed through the open disused gateway outside the walls and then, sweeping round

---

[199] Mr. Crean, who was defending the gateway, told me that "they came on quite splendidly"; he always spoke with the greatest enthusiasm of their courage.

with a cheer of "Avanti Savoia" the historic battle-cry of the Bersaglieri, they dashed up the road. But they had still some 120 yards to cover; the true gate, that is to say, the inlet into the castle yard, still loomed in front of them, under its great stone archway, pierced by their shot, but only partly shaken. And the shout of the Irish had gone up.[200] As the column rushed forward, both Irishmen and Franco-Belgians, some of them standing up recklessly on the walls, poured in a furious fire from the front and flank. A moment later Captain de Baye fired his six-pounder, charged with grape, straight into the center of the column. This checked it for a moment, but then Captain Prevignano called to his men and they came on again splendidly. A lieutenant of the Bersaglieri went down badly wounded, but his stormers reached the gate, and as they hacked at its battered front with their axes, Prevignano's voice could be distinctly heard encouraging them on. Here the real struggle took place: the entrance had been reinforced by a barricade of planks, and "between the beams," says O'Reilly our men answered them with shots and with thrusts of their bayonets."[201] For several minutes it must have been deadly work under that archway. Various incidents are recorded of the hand-to-hand fighting. At one moment Sergeant O'Neill, an old Crimean, threw open a small trap and, putting his

---

[200] The wild cheer of the Irish at this moment seems to have made a surprisingly strong impression on those who heard it. Emmanuel de Gouttepagnon, one of the 23 Franco-Belgians, wrote home an account of the attack, in which he says: "Ils ont ètè bien recu par les Francais et les Irlandais. En cinq minutes nous avons balayé la place par une fusillade bien nourrie et animée. Les Irlandais criaient comme des sauvages; c'était un vacarme épouvantable." ("They were well received by the French and Irish. In five minutes we had swept the place with a lively and well-nourished fusilade. The Irish were shouting like wild Indians, and the noise was terrific.") And Santorelli says: "Ogni scarica era accompagnata da grida feroci degli appiatati stessi, i quali erano pressoche tutti Svizzeri ed Irlandesi." ("Every discharge was accompanied by fierce shouts from those behind cover, who were nearly all Swiss or Irish"): as a matter of fact they were all Irish, except the 23 French. The Swiss were all on the opposite side of the fortress.

[201] O'Reilly's report.

musket through, fired point blank at Prevignano;[202] Prevignano had had his sword smashed off short by a bullet, but he seized the musket with his hand and pushed it aside just in time; whereupon O'Neill instantly thrust at him with the bayonet, but he again succeeded in saving himself by a narrow space. Others were not so fortunate; one of the men with axes was bayoneted by an Irishman through a crevice and fell dead against the gate; and a Piedmontese put his rifle through one of the holes made by the cannon balls and shot Lieutenant Crean in the right biceps.[203] Thus the give and take went on unabated for about five minutes, no man giving way on either side of the planks until suddenly the Piedmontese bugles rang out loud and clear from near the road; they were sounding the "Retire" by order of General Brignone himself, because he saw that the attempt was hopeless. It was then, and only then, that the Bersaglieri at the gate ran back to rejoin the Grenadiers, who had been halted about half-way up the slope. And Prevignano, whose courage had been most conspicuous throughout the whole affair was the last of all to retrace his steps.

As they retired, a second shot from the six-pounder flashed after them out of the gateway, inflicting some further losses, so that the roadway back to the town was all strewn with dead and wounded. This had indeed been the most gallant attempt to storm the place. The whole affair cannot have lasted more than .about a quarter of an hour, or, at the outside, twenty minutes, but in that short space of time the Bersaglieri company had lost at least one-third of its number; during the struggle outside the gate they were falling right and left; yet the remainder had not gone back a foot until the bugle

---

[202] Oddly enough Prevignano was also a veteran of the Crimea. O'Neill was a Limerick man who had seventeen years' service in the British army.
[203] "Lieut. Crean received a ball through the arm while defending the entrance, and distinguished himself together with Captain Coppinger by his courage and coolness."—Major O'Reilly's official report.

The Fight at Spoleto; September 17th

sounded the "Retire."[204]

On the opposite side the casualties had been wonderfully few—several of the Irish had been wounded including Lieutenant Crean, but only one had been killed, a Dublin policeman named Langley. The Franco-Belgians had got off scot-free,[205] although they had exposed themselves recklessly, pouring in their fire on the advancing column.

At the same time, within the walls there had occurred a rather curious incident and one perhaps natural in the case of such very imperfectly trained soldiers.[206] Major O'Reilly, as we have related, was standing near the gateway with two sections of Irish in reserve. They could see nothing, but naturally were very much on the alert at the smallest sound. They heard the bang of the little gun; then the shouts of the stormers, and then suddenly there poured across the courtyard what appeared to be a stampede of Papal soldiers; it was some of the Italian gunners of de Baye who, after firing off their six-pounder, were running back, finding their corner untenable.[207] Such an example is contagious; the boys in the Irish sections concluded, perhaps not unnaturally, that the gate had been forced, and that all was lost; they in their turn broke and ran for the guard house.

On their tracks went Captain Coppinger bent on rallying them, doubtless with some well-chosen expressions on his lips. Some turned back and others halted; but being only recruits they were too much confused to form up again in ranks or

---

204   This Bersaglieri company started the campaign with just-over a hundred men all told, so that there would almost certainly have been less than a hundred taking part in this assault; their losses were 11 killed and 22 wounded.
205   O'Reilly was under the impression that one of them was wounded, but Emmanuel de Gouttepagnon says that none of them were hit so this may have been one of the Irishmen who were with them. O'Reilly mentioned six of them in his report: Sergeant Townley, Privates Crespin, Terrier, Margerie, Baron de Forstner, and Vicomte d'Aigneau.
206   Maj. O'Reilly's diary.
207   Their retirement, according to Dunne, was unavoidable.

to carry out the words of command. Meanwhile, throughout those anxious seconds O'Reilly could hear the axes ringing on the gate, until he thought, as he says, that it must nearly be down; so, abandoning all attempts at re-forming the recruits, he simply jumped off the ledge on which he stood, out into the yard, shouting, "Irishmen, follow me!" This at all events was a plain order, and in a moment the two sections were after him cheering, and dashed into the gateway just as the Piedmontese were retiring.

During the rest of the day General Brignone contented himself with pounding the garrison with artillery fire, and showering the place with rifle bullets. He certainly did not mean to try another assault, for he sent off a messenger to General Fanti at Col Fiorito, about thirty-three miles distant, to say that a regular bombardment of the castle would be necessary; and Fanti gave orders to Colonel Genova di Revel, his staff officer for artillery, to start at once for Spoleto to organize the artillery attack[208] on the following morning.

Nevertheless, although there were no further assaults that afternoon, the firing was kept up with great vigor, the guns aiming at the roof of the keep, and the riflemen at anyone whom they could see. And the defenders had no adequate means of answering them; they could not reach either the riflemen on Monte Luco, or the gunners on the town side; so that although their casualties were few, towards the end of that sultry afternoon they were undoubtedly feeling the strain

---

208   Colonel di Revel says: "When I arrived [at Foligno] on the morning of the 16th I found General Brignone, who with the 3rd Grenadiers, 9th Bersaglieri, two squadrons of Niçois and Ghebart's battery [he twice refers to the battery as Ghebart's— not Duprés, as officially stated] was proceeding to Spoleto. . . ."

On the 17th Colonel di Revel was at Col. Fiorito.

"In the evening the General sent for me and told me to start for Spoleto at once in order to systematize the attack on the Rocca there which Brignone feared he would not be able to force (forzare). Just as I was setting out there arrived a dispatch from Brignone to say that he had occupied Spoleto. . . ."

Da Ancona a Napoli, Miei ricordi, by Genova di Revel, p. 39.

The Fight at Spoleto; September 17th

of being shelled for so many hours without being able to reply. And more than anything they felt the thirst. "My mouth was parched," said one of them to me; "not merely from the heat, but from biting the cartridges,"[209]—and it will be seen that O'Reilly in his diary gives a vivid account of their recklessness in filling their canteens under fire. But their most serious danger was due to shells falling on the keep in which was situated the powder magazine. On several occasions during the course of that afternoon the roof of the keep was actually set on fire, though only once did the matter become serious; on that occasion O'Reilly gives the following account:

> I at once went up to the keep with Sergeant Lloyd, and as I passed up desired one or two of the Swiss officers to take fifty of their men and come to put it out. It will give some idea of the closeness of the fire of the Bersaglieri on the mountain whenever anyone was seen walking up to the keep, to mention that whilst mounting the hill, from the small breach for new work to the end of the guard house where one was sheltered for a minute, at the same moment a bullet went close over my head besides a dozen others that went wide. On leaving the shelter of the gable of the guard house one was still more exposed both to rifles and to the chance of a round shot. Just after I passed, a round shot banged against the jamb of the gate.

It was under this fire, however, that the Swiss recruits arrived and extinguished the blazing roof; and after that evening closed in without further mishap.

At about 8 p.m. the firing ceased. Certainly as far as the fighting was concerned the volunteers had every reason to

---
209   Dunne.

be satisfied. Though few of them were professional soldiers, though hardly any of them possessed their full kit, and several of them were still in the rough frieze coats in which they had left Ireland, they had taken a grip of their old courtyard wall and were still holding it as firmly as ever after twelve hours of artillery fire: and the twenty-three Franco-Belgians, the dozen or so Austrians and Bavarians, had stood the same test, while the untrained Swiss recruits had done astonishingly well.

Yet, in spite of his success, O'Reilly was considering the question of coming to terms, and one sees very plainly the difficulty of his position.

In the first place there was no question of victory. Victory had never been possible; and even the people for whom he was risking his life were in opposition to him. Secondly, there was no question of his obtaining any advantage for his side by prolonging the defense; in this respect his position was entirely different from that of General Schmidt who had been left at Perugia with 1,650 regular troops in order to delay the march of Fanti's army; and thirdly, he had been definitely informed that no attempt would ever be made to relieve him. So he was, in fact, fighting solely for honor.

At the close of that evening he was—as the German writer Rustow has pointed out—at the zenith of his possible defense. He had won an initial success. It was evident that General Brignone had found the old circle of walls a harder nut to crack than he had expected, and he would doubtless be ready to agree to honorable terms. But O'Reilly was quite aware that his own success could not possibly continue; as matters stood the defense might not survive the night, and in any case it would hardly last beyond the following evening.

His casualties had been wonderfully few;[210] but the fighting

---

210 V. List of Irish casualties at the end of the book. O'Reilly reported only three men killed and 10 wounded, but it seems probable that there were about about 17 or 18 in all. Dunne told me that some men did not report themselves wounded because they were afraid of being left behind in Italy. They had no confi-

portion of his garrison was tired out, having: been under fire all day and awake the night before, and hard at work the three previous days; and he had no one to put in their place on the wall. His reserve most certainly could not be relied on.[211] In such a case an assault by night, if carried out with the same reckless dash and self-sacrifice that had been shown by day, might again throw confusion among his recruits, and would have quite a good chance of penetrating the defenses, the more so as the unfinished portion of the wall was practicable for escalade. But apart from night attacks, it seemed unlikely that the place would be tenable by the following evening. His supply of rifle cartridges was running short, and once that they were ended he would have no means of preventing the Piedmontese from pushing forward their guns and pounding his walls at two hundred and fifty yards' range if they liked.

He decided, therefore—after consultation with Captain Coppinger and Captain de Baye—that it was better to try and get the most suitable terms possible as a result of their day's work, than to risk a "mop-up" during the night or accept the certainty of receiving worse terms on the following evening. With this end in view he asked for a truce for the purpose of carrying away the wounded, and at the same time he allowed the Papal Delegate to go down to General Brignone—which he had already expressed his wish to do—and discover whether it would be possible to arrange for honorable terms of surrender. This mission proved successful, and shortly afterwards O'Reilly himself went down accompanied by Captain de Baye.

General Brignone received him with great courtesy, and expressed himself willing to concede anything that his orders

---

dence whatever in the Piedmontese (*cf.* the case of Clooney's party in Perugia).
Dunne said, "there would be hardly 20 in all."
The Piedmontese losses are stated at 14 killed and 49 wounded. (Brignone's report) or at 16 killed and 48 wounded (Brignone's letter of Sept. 21st to Fanti.) *V.* Appendix G.
211   This fact is not questioned on either side.

allowed. While O'Reilly on his part said quite openly that although he could hold the place for some time he foresaw that he must ultimately surrender it and was therefore prepared to yield if offered honorable terms. So after some discussion it was agreed that the garrison should capitulate; the men were to surrender their arms and march to the confines of the state, after which they were free to return to their own homes; the officers to retain their swords, but to give their parole not to serve again during the campaign. This latter clause gave rise to a rather graceful little episode. O'Reilly had raised an objection to the officers giving their parole, so General Brignone assured him that his orders left him no choice and that Schmidt had agreed to this concession at Perugia; but with a kindly thought which has never been forgotten among the volunteers, he instantly turned to his A.D.C. and dictated the following passage: "Gli ufficiali e le truppe saranno ovunque trattati con quella urbanita che si addice a truppe onorate e valorose come hanno dimostrato di esserlo nel combattimento di oggi,"[212] which means: "the officers and men will everywhere be treated with that courtesy which is due to troops that are honorable and brave as they have shown themselves to be during the fighting today."

This, then, was the end of Companies Nos. 2 and 3 of the Battalion of St. Patrick as such—although individually nearly all the officers rejoined the Papal service as soon as their parole expired, and so did some of the men.

But the fight at Spoleto had at all events one satisfactory result not always to be found on such occasions: that the combatants on either side parted with a better understanding

---

[212] This is the version given in the Italian official history of the campaign. O'Reilly in his diary gives a rather different wording, conveying the same meaning, and adds that General Brignone then turned to him and said: "It is only the truth, and I feel bound to state it." The insertion of this clause was evidently solely due to the courtesy of General Brignone, as there is nothing similar in any of the other agreements of surrender.

of each other. Fifty years later, when the Irish veterans spoke of the assault, even the most Papal of them never failed to repeat their high admiration of the courage shown during that small but gallant episode. "The Piedmontese came on splendidly," seems to represent their memory of the scene.[213]

During those fifty years Italy has become a great nation; but even now in 1924, when the Italian Headquarters published their official account of this early campaign, the first in which their national army recorded its victories, they have expressed their verdict about the Spoleto episode in the following terms:—

> From a military point of view the conduct of Major O'Reilly is to be commended: he did his duty, well supported by his Irish and by the Franco-Belgians, and General Brignone made a chivalrous recognition of this fact by designating as honorable and brave troops all those who had that day fought against him; therefore it is certain that if the Papal commander had not been obliged to improvise his resistance, and if he had been in command, not of an omnium gatherum of different details and soldiers collected hastily and at random, but of a properly appointed unit and one well known to him, he could have prolonged the resistance. In this episode, then, we may find yet another result of General de La Moricière's having been taken unawares in the field of strategy.[214]

---

213 "It was impossible to repress a feeling of sympathy and admiration for the brave men who in the face of a deadly fire, without cover or shelter, went on without swerving." Article by Mr. M. T. Crean (the officer posted in the gate), published in the *Seven Hills Magazine* of March, 1908.
214   M.G. Uff. Stor. (Italian official history), p. 261. La campagna delle Marche e dell' Umbria.

CHAPTER XVII.
No. 4 Company's March to Castelfidardo.

We must now leave Nos. 2 and 3 Companies on their march to Genoa, the port at which they were to end their eventful three-months' tour in Italy, and return to follow the fortunes of No. 4 Company, the only Irish unit that had been selected to march with de La Moricière's field force; we last left it starting from Foligno on the evening of September 12 in a state of great enthusiasm, amid the cheers of its companions. Of this company, unfortunately, no actual record exists, and I have been unable to find a single survivor belonging to it—probably because it was so small. It consisted most likely of about 105 men[215] and three officers: Captain Kirwan, a keen, hard-working officer, with some previous experience in the Militia; Lieutenant Carey, also an ex-militia officer; and the young 2nd-Lieutenant D'Arcy, a boy not yet seventeen years of age,[216] who seems to have been a born soldier.

It will be unnecessary here to give any detailed description of that lightning march northwards of de La Moricière, because we hear of no incidents that especially concern the Irish company; they were in charge of the artillery, and that is all that we know about them, except that they were short of

---
215 De La Moricière had originally intended to take 100 Irish men, but there was great competition to go, and certainly a greater number finally went.
216 Born 13th March, 1844, according to the Matricola in the Royal Archives.

some of their kit. The Papal stores, as already mentioned, had run out of almost everything, and it was due to this fact that our volunteers had no pouches and were therefore compelled to carry large bags, in which were stuffed both their cartridges and their rations; and many of them must have been short of other things as well.

The Papal field force was divided into two portions, first, the leading half, consisting of Cropt's Brigade, about 3,000 strong, of which de La Moricière took command in person, and secondly, Pimodan's Brigade, about 3,500 strong, which followed a day's march in rear.

Starting on September 12 from Foligno, at about seven o'clock in the evening, de La Moricière marched north-eastwards over the Apennines and right across the Papal State to the town of Loreto, with only three halts—Foligno, Serravalle, Tolentino, Macerata, Loreto—followed at one day in rear by Pimodan. On the evening of the 16th, he arrived at Loreto, having accomplished 75 miles in four days in such sultry weather that even the Italians were exhausted, and most of the marching had to be done by night. It was with his force that the Irish company was included. In their rear, Pimodan's Brigade, starting a day earlier from Temi and reaching Loreto only a day later, had accomplished 106 miles in six days, arriving late in the afternoon of the 17th, the evening before the battle—after covering twenty-two miles. De La Moricière had been halted all day at Loreto so as to allow Pimodan to rejoin him, for he saw that he would not be able to get through to Ancona without a fight.

By this rapid movement de La Moricière had outstripped the southern Piedmontese divisions of Fanti and Della Rocca, who were following him after their capture of Perugia;[217]

---

217   On the 17th, as Fanti's column was winding its way among the mountains around Colfiorito "between the hours of 3 and 5 in the afternoon, it heard distinctly the echo of the guns down in the valley, showing that the pontifical resistance around the Rocca at Spoleto was surpassing what had been expected."—M.G.

but he had failed to get northwards in time to slip through into Ancona before being intercepted by Cialdini, who, with 17,000 men, had dashed down from the north and was established in front of him, barring his way.[218] Their two opposing forces were now very close to one another. Loreto stands on a height and, from that height de La Moricière could look across the valley to where, about two miles distant, Cialdini had lined out his men on the hills of Castelfidardo, behind the river Musone. De La Moricière would have to cross that river and force a passage over or round those hills before he could reach his haven of refuge, the fortified town of Ancona, which lay about fourteen miles to the far side of Cialdini. In the meanwhile, he had marched his recruits to a standstill. The only point in his favor was that Cialdini's men had been almost as hard pressed. But—and this was the most unlucky chance of all for the Papal side—whereas both Cialdini and de La Moricière had been able to give their men a day's rest, Pimodan's unfortunate brigade, that is to say, the best armed—in fact, the fighting portion of the Papal army—being twenty-four hours in the rear of the others, had only arrived at Loreto on the evening before the battle, fasting and tired out, after a march of twenty-two miles, at the end of nearly a week of physical and mental strain continuing by night as well as by day.

It is a profound pity that we have no record of this march from any of the Irishmen who took part in it, because it must have been a wonderful experience for these boys, especially during the night of the 12th and 13th, when they wound their way by zig-zag mountain roads across the Apennines, and slept at Serravalle in the snow. We know little about them except that they certainly had a very harassing time, being

---

Uff. Stor., p. 251. (Italian official history of the campaign). This was a remarkable distance for sound to travel, nearly 30 miles.
218   V. Vigevano Also official reports of De La M., of Fanti, Cialdini, etc. and many books of memoirs.

in charge of the artillery and of the wagons containing the treasury's supply of specie for the whole army, amounting to about a million lire. This must have added greatly to their trials. Moreover, the guns were so imperfectly horsed that spans of oxen had to be yoked to the teams to get them across the steep or damaged stretches of road, and after the first day they were allowed to fall back in the rear of the column. All this must have greatly increased the already wearing duty of the men with them, besides lengthening the hours which they were compelled to spend on the road.[219]

One small detail we have as to the feelings of the Irish volunteers during these days, namely, the deep impression made upon some of them when they visited the shrine at Loreto. This can be understood when one remembers that they were seeing it for the first time, on the eve of their first battle, and that the whole army profoundly realized how small were their chances of success. Count Becdelièvre, the colonel of the Franco-Belgian Tirailleurs, had addressed his 270 men in a short speech of which various versions have been printed, but the following is his own:—

> Tomorrow, at this hour, several among us will have appeared before God. Now, you know that a man should be (clean when he appears before Him; let those who' are not so go round to the office to our chaplain. I have only just left him myself.

---

[219] The details of this march are taken from the Italian account of the campaign. *Vide* M.G. Uff. Stor., pp. 2, 4, etc. There are many descriptions of it, mostly by Frenchmen.

CHAPTER XVII.

THE BATTLE OF CASTELFIDARDO; SEPTEMBER 18TH.[220]

To give a brief sketch of the battle of Castelfidardo: strategically de La Moricière was in a desperate position; he could not go back and he could not go forward; tactically, however, he showed all his traditional skill, and, with soldiers such as those whom he had led in his younger days he would probably have extricated himself from his difficulty, but, as matters stood, his untrained men "let him down."

His plan may be detailed as follows: the Piedmontese (facing south) had occupied the line of hills opposite him, behind the river Musone. But the end of this line of hills finishes off about five miles inland from the Adriatic, so that there was a very wide gap between the end of the Piedmontese left flank and the sea. De La Moricière was determined to push through this opening—which is a flat, fertile plain—and make his way along the coast to Ancona. Cialdini of course was perfectly well aware of this gap, but he believed that in its lower regions the River Musone was too broad and too deep to be crossed

---

[220] It is impossible to give more than a few of the authorities on the battle of Castelfidardo. The best modern accounts are those of Vigevano and Barbarich (both official). But excellent narratives are to be found in all the military reports. That of de La Moricière's is the best. Fanti's is also good; and Cialdini's of course is necessary. Many of the minor reports have been published by Barbarich (Castelfidardo); but some have still to be sought in the Archives.

by an army.

As a matter of fact there existed two fords through it: one was at Casa Arenici just outside the end of the hills, and Cialdini had ordered a Bersaglieri company to guard it—to be detached from the 26th Bersaglieri battalion which was posted on the outer edge of the Montoro height, the extreme left-hand point of his line.

But there was also another ford lower down, situated at Casa Camiletti, about two and a half miles outside the left of Cialdini's line and of this he knew nothing. Yet an active enemy crossing at this Casa Camiletti ford could march past the left flank of the Piedmontese and get into Ancona via the coast road to Umana—a fact of which de La Moricière was perfectly well aware.

So with an audacity characteristic of his African days, de La Moricière had formed a plan for launching his little army in two long columns of route, each directed at one of the fords, and marching on parallel roads, so as to cross more or less simultaneously. His left-hand column was to wade through the river at Casa Arenici, and the leading battalions were to attack and brush aside the Bersaglieri there, and thus get across Cialdini's left flank, while the right column, with the baggage train, would cross unopposed at Casa Camiletti in perfect safety, over two miles to the right and completely outside the field of action.

By this means he would have placed his whole 6,500 men right across Cialdini's left flank, in two long columns. These columns would then face inwards to their left, and try and roll up Cialdini's left until they had captured the end hill. They could defend this end hill (Montoro) while their baggage got safely away on the road to Ancona, and then they could retire and follow in its tracks northwards.

The touch of genius in this plan lay in the fact that it enabled him to throw the whole of his little force—arrayed in two lines

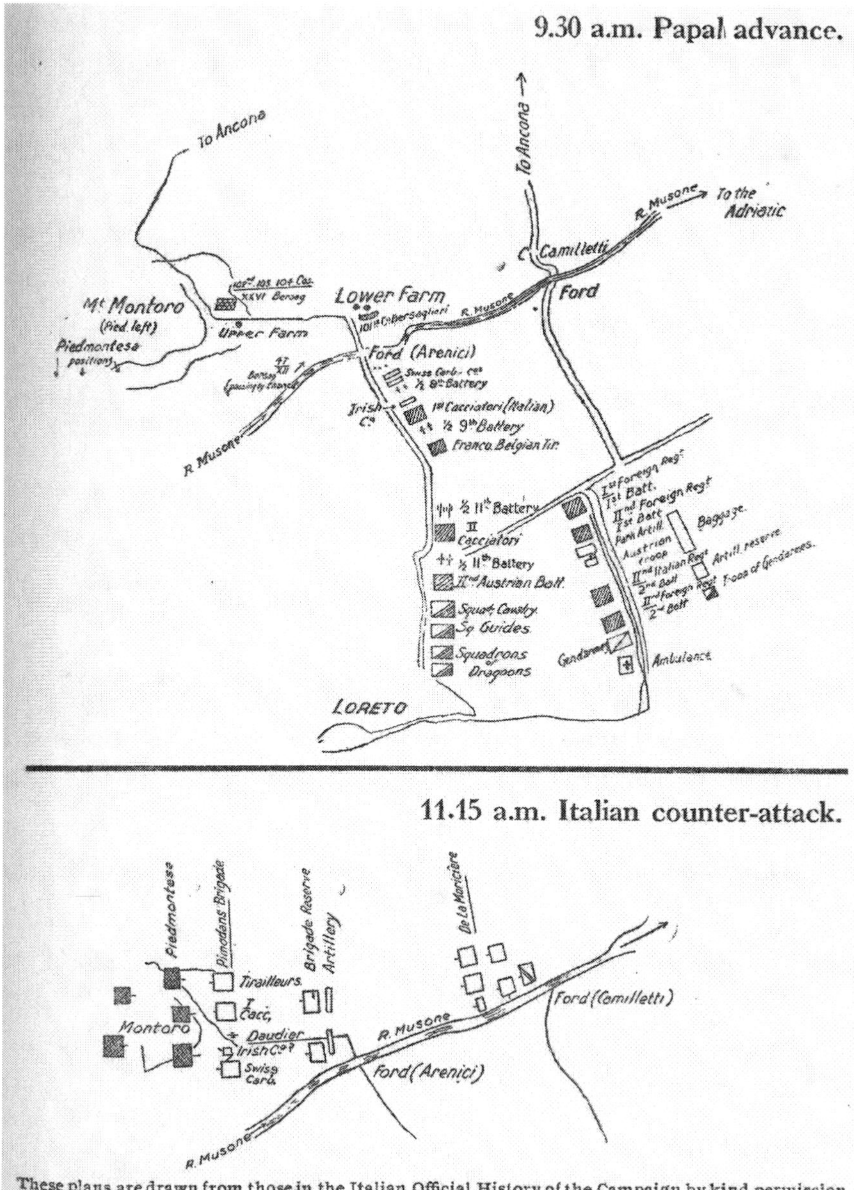

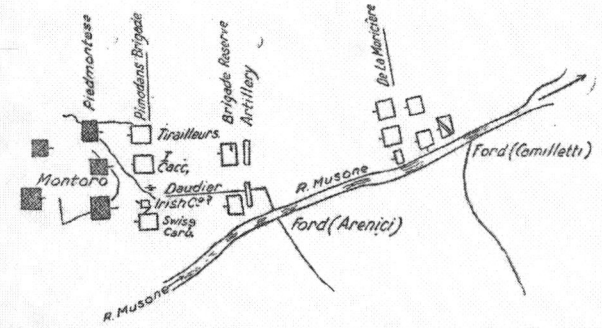

These plans are drawn from those in the Italian Official History of the Campaign by kind permission of the Historical Section of the Headquarters Staff (Stato Maggiore Centrale.—*Ufficio Storico*).

(Scan curtesy of Ave Maria Canizaro Library)

The Battle of Castelfidardo; September 18th

of attack one behind the other—against the left-hand point of Cialdini's line, where it was weakly defended.

We will begin with a short description of the battle.

His scheme nearly succeeded. The left-hand column, consisting of Pimodan's men, drove back the Bersaglieri at the Arenici ford, got across their flank, and faced inwards to its left. It then attacked the hill of Montoro and won its way three-quarters of the distance to the summit. Meanwhile the right-hand column had crossed at the far-off Casa Camiletti ford, and also turned inwards according to plan, thus taking up the required position as second line and reserve of the attacking force. But when de La Moricière sent to this second line the order to advance and support the front line their nerve failed; not one of them would come under fire.

Consequently the Piedmontese reinforcements, arriving rapidly on the top of Montoro, were able to charge headlong down its side, and after a tremendous struggle utterly destroyed Pimodan's brigade, and drove its scattered remnants into the plain amid fearful confusion. The final result was that about 3,000 Papal soldiers of all units and nationalities sought refuge in Loreto, nearly 3,000 fled far and wide, and de La Moricière, rather than be included in the inevitable surrender, galloped away with a few followers to Ancona to direct the defense there.

In this engagement, as in the others, the Papal army was divided into two distinct sections; those who fought and those who failed to do so. Those who fought consisted of the Franco-Belgian half-battalion, the Austrian battalion (Major Fuchmann), one of the Swiss battalions (Colonel Jeannerat), the Irish company, and some of the Italians. Pimodan, who led them with the most sublime gallantry, was wounded three times and died on the same evening.

*   *   *   *   *   *

Having begun by giving the above skeleton description of the battle we will next enter upon its detail in so far as they concern our subject. Amid the supreme confusion of this defeat it is very difficult to discover exactly what happened to the Irish company. There are various reports from colonels and other senior officers, each glorifying his own battalion and decrying everyone else, but there is nothing from any of the three Irish officers; their unit was so small that it sent in no report, and consequently it is at a disadvantage, or rather at the mercy, of the somewhat self-interested statements of others.

We know that the Irish company was included in Pimodan's attacking force, although hitherto it had been part of de La Moricière's column; it was the only infantry unit thus transferred from the one to the other, but this was probably due to the fact that it went with its battery. The Irish company was also. we may note, the only unit in the attacking brigade armed with the old smooth-bore muskets—and this was a very great disadvantage against the Piedmontese rifles.

According to the modem Italian official accounts, the Papal troops all started fasting. This seems to have been partly due to bad management, and partly inevitable, but coming after four or five days' marching it accounted most probably for their complete breakdown a few hours later.

Pimodan's column moved off from the hill of Loreto at 8.30 a.m on its way to attack the Bersaglieri at the Casa Arenici ford; de La Moricière's right-hand column with the baggage train did not start until nine o'clock on its way to cross the Casa Camiletti ford, over two miles to the east.

At the head of Pimodan's column were the Swiss Carbineers, and next in rear of them came half the 9th Battery; then the Irish company, and behind them the 1st Cacciatori (Italian); then the second half of the 9th Battery; and following them the half-battalion of Franco-Belgian Tirailleurs. These were to form the front line.

About five hundred yards in rear came the *brigade* reserve, consisting of the 11th Battery, of the 2nd Cacciatori (Italian) and of the 2nd Austrian battalion[221] (Fuchmann's) with some cavalry.

By about 9.30 this (Pimodan's) column had wound its way down into the plain and was close to the Arenici ford when its advanced guard ran into and apparently surprised an outpost of the Bersaglieri on the near side of the river. The Swiss Carbineers at once drove it out and pursued it across the river, and were very soon engaged with two Bersaglieri companies on the far side.[222] Meanwhile the Irish splashed through the ford after them, in water about two-feet-six deep, accompanying their guns, and were shortly followed by the other battalions. At the end of half an hour all Pimodan's front line was over. Nevertheless the Bersaglieri, though outnumbered, had fought very cleverly in open order, skirmishing among the reeds which covered the river side, and winning invaluable time for their reinforcements to arrive on the scene.

The next task before Pimodan—at about ten o'clock—was to take the Lower Farm, the Casa di Sotto.[223] This was defended by five Bersaglieri companies: namely, the two originally there, and the remaining three of the 26th Battalion (Bersaglieri) which had now come down from the end of the hill to reinforce those below. These totaled about 550 men in all, but, of course, they were far outnumbered by the attacking units which consisted of the six Swiss companies (on the left), the 1st Italian Cacciatori (center), and the

---

221   Officially called the 2nd Bersaglieri; but it is best not to call them by that name, because it causes them to be confused with the Piedmontese Bersaglieri.

222   One of these was the 101st, the company told off to defend the ford; the other was the 47th which happened to be passing on its way to relieve another company.

223   Also called the Casa Andreani-Catena. This was a farm and some few cottages standing in the plain just opposite the Arenici ford, about two or three hundred yards from it.

Battle of Castelfidardo; the River Musone at the point where the Irish crossed it (from left to right).

(Scan curtesy of Ave Maria Canizaro Library)

The Battle of Castelfidardo; September 18th 187

Franco-Belgian half-battalion of Tirailleurs (right); with the guns in rear escorted by the Irish—about 1,800 men in all.[224] And the attack would certainly have had a rapid success but for the fact that when the Swiss were advancing they were fired on from behind by the 2nd Cacciatori, which was in reserve[225]—a contretemps which checked their advance and wasted some invaluable time. When at length the attack was resumed it was very skillfully met by the Bersaglieri: they played their game well; defended, counterattacked, wasted time, and finally slipped away to their right up the hill down which they had come. Over half-an-hour had been spent in winning about three hundred yards.

It was during this period that the Irish had their first experience of bloodshed. They were rather in rear with their guns, where the cart-track leads out of the ford, and the bunches of grapes hang in festoons between the short elm trees, when suddenly they saw a line of athletic Bersaglieri—unmistakable in their feather-plumed hats—dashing on to them at the charge. This daring attempt was due to the Bersaglieri officer, Captain Nullo, who had seen a chance of capturing the artillery section and had worked his way round the Bank of the firing line and had then dashed through the gap headlong. It was an effort that deserved better success than it achieved. Nullo led two rushes with reckless abandon, and even came to bayonet-fighting with the Irish and the gunners, but on the second occasion he fell mortally wounded and was made prisoner, together with one of his subalterns, a sergeant, and some of his men. At first these prisoners remained in the hands of their captors, but soon the other Piedmontese subaltern, Lieutenant Canina, came to their assistance, and—so he

---

224   The Swiss Carbineers were less than 600 strong, two of their companies being in garrison. The 1st Cacciatori were a little over 700 strong, and the Franco-Beiges about 270.
225   This was at first recorded as an act of treachery, but it is now attributed merely to unsteadiness.

tells us in his report—leading the left half-company twice to the charge, he succeeded in rescuing his dying captain and those immediately round him. The Bersaglieri then retired, leaving two of their men still prisoners, and having suffered probably from twelve to fifteen casualties during the tussle.[226]

*   *   *   *   *   *

At 10.30 Pimodan had taken the Lower Farm (Casa Andreani-Catena), and the Bersaglieri had retired half-way up the hill to the Upper Farm (Casa Serenella); so the Papal troops were now across the Piedmontese flank, and they formed front to their left, according to plan, in order to attack the Montoro hill and establish themselves on top of it until their convoy should be safely through on its way to Ancona. In this attack the same order was preserved as in the first—the Swiss on the left, the 1st Cacciatori in the center, and the Franco-Belgians on the right; but, according to the Italian official narrative, the Irish company seems to have advanced on the right of the Swiss with Daudier's howitzers a little in rear of it.[227] De La Moricière's report, as will be seen, speaks shortly afterwards of the Irish dragging one of Daudier's howitzers up the hill under fire, and then joining on to the Franco-Belgians, so they had evidently remained with the battery.

The brigade reserve consisting of Fuchmann's Austrian battalion and the 2nd Cacciatori (Italians) with its cavalry and artillery, had by now crossed the river and halted near

---

226   This Bersaglieri company had 19 casualties during die day, but this was by far its closest engagement.

227   "In order to give more cohesion and direction to the impetuous but rather confused advance of his men, General Pimodan succeeded in stopping it for a moment, and formed a column of attack as follows—the battalion of Swiss Carbineers and the Irish company to the left, in the center the 1st Battalion of Cacciatori, and on the right the Franco-Belgian Tirailleurs."—M.G. Uff. Stor., p. 345.

It looks rather as if the Franco-Belgians and Irish company had come nearer to one another during the course of the attack: but one cannot be certain as to exactly how this happened.

the ford.

Pimodan led the attack, and at the end of twenty minutes he had taken the Upper Farm (Casa Serenella), but not without loss. And the Bersaglieri, although they had fired away nearly all their cartridges, merely retired up the hill towards the Casa Corraini, a little further back on their left rear, without abandoning the fight. They had now been driven out of three different positions but nevertheless—though completely outnumbered—they had accomplished their mission to the full. They had delayed the Papal attack for an hour and twenty minutes, and gained time for the Piedmontese reinforcements to arrive. At that very moment the four battalions of the 10th Regiment were hurrying along the ridge of hills from Crocette to reinforce the Piedmontese left on Montoro; a battery, consisting partly of rifled guns which would outrange those of the Papalini down below, was coming with them, and General Cialdini himself was galloping along to the end of his line to take charge of the situation.

At 10.50, therefore, when Pimodan started to lead his culminating assault against the summit of the hill—which is a steep incline with some trees on it, and crowned by the house called the Casina Sciava—the battle had resolved itself into little else than a race for the summit. If the Papal troops arrived at the top unchecked they might very likely be able to hold out there for as long as was necessary, but if the Piedmontese occupied it first they would be very difficult to dislodge.

It was a near thing. On the Papal side the Swiss, the Cacciatori and Franco-Belgians had actually reached a line just below the road that runs along the hillside, when the first two battalions of the 10th Piedmontese Infantry Regiment arrived at a double and took up their positions on the summit. As they were doing so, Cialdini and his staff came cantering along; and at that moment so near to them were the Papal troops below, that these staff officers could plainly distinguish their

French words of command. Cialdini at once gave the order to ground knapsacks and charge, and his two infantry battalions, about 800 men in all, hurled themselves over the edge to the rescue of the Bersaglieri. Thus reinforced, the Piedmontese in the fighting line were now about 1,200 strong, all regular troops, and most of them fresh for the fray; the Papal infantry there must still have numbered between 1,600 and 1,700, but they were only half-trained men who had been fighting for an hour and a half and had had nothing to eat that day.

The result was a tremendous melee; the Italian line came down like a torrent marked out at many points by the plumed hats of the Bersaglieri. But it was well met. The Swiss Carbineers stood firm, and the Franco-Belgian Tirailleurs were beyond all praise; and Pimodan rode fearlessly among them, already twice wounded, a mark for every bullet that flew. Presently he launched a counter-attack with the bayonet, which was carried out with indescribable enthusiasm, especially by the young Franco-Belgians; and the Piedmontese were driven back nearly two hundred yards. But their other two battalions of the 10th Regiment had now arrived on the summit, and another 800 fresh men were hurled over the brow of the hill. The Papal line, outnumbered for the first time, drew back a little, but rallied in the most gallant fashion and began a desperate resistance round the Upper Farm. And—at this supreme moment, when everything was still to be won or lost—de La Moricière ordered Pimodan's brigade-reserve, about 1,500 strong, to come up from the Arenici ford and reinforce the firing-line; while, further back, he sent for his own second line of 3,000 men from the more distant ford of Casa Camiletti to come forward and occupy the place of the brigade-reserve.

On this reinforcement depended the whole fate of the Papal army. The attacking battalions were halfway up a steep hill, clinging desperately to their ground against rather superior

The Battle of Castelfidardo; September 18th 191

numbers, but one determined rush by reinforcements might perhaps have swept them right up to the summit of the position, or at all events have gained a respite, for their retirement. At that moment, however, the second line could not be persuaded to advance. Without having a single man killed it broke and fled; and this shameful example very soon spread panic even in Pimodan's brigade-reserve, which had moved forward to the foot of the hill. When it received the order to reinforce the firing-line, Fuchmann's Austrian battalion advanced as coolly as on parade, but the 2nd Cacciatori fled; as did some of the cavalry; and so did the gunners of the 11th Battery, whose pieces were stuck in the mud.

The result was that the firing-line, abandoned by its supports and reserves, was left alone on the side of the hill to face its ruin. For a while it resisted desperately, but then was gradually compelled to retire step by step as it had mounted. Firstly, it clung to the Upper Farm, then to the Lower Farm, and finally wended its way back to Loreto. The struggle at the Upper Farm was a scene of frightful confusion, of close hand-to-hand fighting, until the defense was broken and the building blazing to the sky. Then the Papal troops, now all intermingled, retired to the Lower Farm, and the day ended up with a last melee there, in a hopeless endeavor to defend their wounded general, their colors and their guns. Finally they extricated themselves, and, covered by the Austrian battalion, followed in the tracks of their second line, which had already got back to Loreto. One notices that they were not pursued.

On the following day the whole force at Loreto decided to surrender. In the absence of de La Moricière it was under the command of General Goudenhoven, and his reasons for surrendering were similar to those given by General Schmidt at Perugia—that most of his men would not fight.

During these last tumultuous phases of the battle it is very hard to discover how matters fared with our Irish volunteers; still, among the seething waves and wreckage around, from time to time they come to the surface, and at one point they certainly earned a very fine mention from their general. This occurred after the first attack on the summit had been repulsed. The Papal artillery was making desperate efforts to support its infantry, and Lieutenant Daudier de Laval tried hard to advance his section of howitzers up the hill as far as the level of the Upper Farm. The distance was about 220 yards over a fairly steep and soft road with ploughed fields on either side, and partly under fire—so much so, at all events, that, according to some accounts, many of Daudier's gunners had deserted him. However, the Irish succeeded in hauling one of the howitzers up to the top and were rewarded by the following record of their action, embodied in de La Moricière's official report to his government:—

> These brave soldiers, after having accomplished the mission entrusted to them, joined themselves to the Tirailleurs [the Franco-Belgians], and during the rest of the battle distinguished themselves in their midst.[228]

It was worth while coming all the way from Ireland to get that mention.

By the time that they emerged on to the halfway level with their howitzer they must-have been very near the enemy, and it is evident that they passed at once into the firing-line. At that moment all around them was a scene of desperate confusion on the hillside. The Franco-Belgians—and among them espe-

---

[228] *De La Moricière's Report,* p. 27. This report was published immediately after the campaign. De La Moricière's it must be remembered was speaking of what he had actually seen, as he had ridden forward to direct the firing-line. Edward Dunne had known one of the Castelfidardo men, and his friend related to him how at one moment the General was with the Irish company, and modified their rather too rapid advance by calling to them "Irlandais," and signaling to them with a wave of his cane to keep steady; he knew no English. Dunne reproduced the French word and the unmistakably French gesture with wonderful fidelity.

The Battle of Castelfidardo; September 18th

cially the Bretons—were sacrificing their lives with absolutely reckless gallantry, charging again and again, and Jeannerat's Swiss Carbineers were making a fine defense. The Irish, with their old muskets, must have been at a great disadvantage, but when the word was given for the line to charge, according to all accounts they dashed in headlong. And the man who went first was their young sixteen years-old 2nd Lieutenant, D'Arcy. His gallantry was noticed by the staff-officers, and he won his Cross of St. Gregory that day.[229] Of the other battalions, the Italian Cacciatori held their position well, and Fuchmann's Austrians did splendidly. But the remainder of the infantry had already departed before the retirement began.

During the retirement it is hard to discover details about the Irish company, but the following quotations from an official description of the battle published by the Italian War Office in the year 1903 gives an idea of the defense of the Upper Farm:

> More bitter and furious was the struggle round the Upper Farm. Carbineers and Tirailleurs (Franco-Belgians) were clinging to it, encouraged m their resistance by General Pimodan, by Major Becdelièvre and by Captain Charette. One cannon (Lieutenant Daudier) laboriously dragged up there by hand and arm work of the Irish of St. Patrick's battalion, had reinvigorated and consolidated the defense. Bersaglieri and Linesmen of the 10th Regiment had surrounded the house; the flames from the straw on fire rose high and smoking into the air and obscured all

---

[229] Count Russell of Killough spent some hours with de La Moricière's two A.D.C's that same evening after the battle, and they told him that D'Arcy had led the movement followed by about eighty Irish. Allowing for casualties and for a few men left with them, this would account for the whole company. The staff officers spoke of "eighty Irish," no doubt because they thought that the company numbered 140 men like every other Papal company, but it had been reduced by about 30 before starting. Of the officers D'Arcy was evidently the most dashing.

view of individual actions.[230]

Later, speaking of the final melee round the Lower Farm, the same account says:

> Individual groups of gunners, Zouaves (Franco-Belgians) Carbineers, Cacciatori, and Irish engaged in a tremendous melee with our men.[231]

And after the fighting had ceased, their final retirement is thus described in the Italian official history of the campaign, published in 1924:

> Carbineers, Irish, men of the 1st Battalion of Cacciatori, and Franco-Belgian Tirailleurs formed the main body and marched intermingled, carrying some of their wounded and dragging with them along the roads seven out of the twelve pieces with which they had started in the morning (three had been taken by the enemy, and the two commanded by Lieutenant Uhde had retired elsewhere), and all the caissons except four; in rear of them was the troop of Guides of the Vicomte de Saintenac which had succeeded in isolating itself when the orderlies, dragoons, and gendarmes were disbanding in confusion, and last of all came Fuchmann's Austrian battalion. At about 2 p.m. these troops reached Loreto and at once prepared for defense, throwing out an irregular line of outposts along the brow of the height,[232] they still preserved an appearance of cohesion, and

---
230   Barbarich La battaglia di Castelfidardo (official narrative), p. 28.
231   Barbarich Ib. p. 31.
232   The O'Clery says that one of the gates of Loreto was allotted to the Irish for defense.

The Battle of Castelfidardo; September 18th 195

halted there although they were only a mile and three-quarters from the ground on which they had fought.[233]

On the following day Count Goudenhoven, an Austrian, surrendered, and gives his reason in the following terms:—

> With the help of the Intendente Gagliani, who was most energetic m dealing with the question of supplies, with the Irish, with the few remaining Franco-Belgians, with the Guides who were still in their ranks and ready to fight, and above all, with Fuchmann's battalion, I might have continued the defense of Loreto. But, in other corps, morale had become weak and they would have impeded rather than assisted us.

Some 3,094 Papal soldiers surrendered at Loreto, and about 2,500 were already scattered far and wide, trying to find their homes; most of the latter were intercepted and made prisoners by Fanti's army as it marched northwards.

De La Moricière, however, had very rightly decided to gallop through to Ancona, well knowing that without his presence there the defense would not last long.

The losses in this engagement seem extraordinarily small compared with the total numbers on either side. But they are not so small if one considers only the troops which were actually engaged. Of the Piedmontese army, only about 2,400 were really engaged at Montoro, and it is very hard to believe the assertion of Fanti's returns that the total number of their wounded amounted only to something over 140, especially in view of the fact that they admit to sixty-two killed. At that rate their total number of casualties would hardly have surpassed

---

[233] Vigevano La campagne delle Marche e dell' Umbria (official history) p. 358.

200. On the Papal side the Franco-Belgians went into action about 270 strong and had about 145 casualties of whom 25 or 30 were killed; the Swiss Carbineers started the day numbering less than 600 and, according to their colonel, lost about 200 men; of the casualties of the Cacciatori battalion nothing is known except that they lost six officers, of whom two were killed; from the Irish company there is no return, but its losses were "heavy," says the Italian official narrative;[234] unfortunately it is impossible even to estimate them, because we do not know its exact strength at the; beginning of the day, and we cannot tell how many Irishmen joined it in the evening from other corps.

The total losses on the Papal side were given by the Pied-

---

[234] The only indication as to the Irish losses that I have ever been able to find came from Mr. W. P. Ryan, the former President of the Irish Literary Society. He was an old Papal Zouave, and had known one of the Irishmen who fought at Castelfidardo. He told me that his friend said to him, "they had between 20 and 30 wounded that day." No doubt the friend intended this total to include all casualties. It is a likely enough figure, but 20 would probably be nearer the mark than 30, because after the battle 103 Irish soldiers turned up at Recanati as prisoners. Some of whom no doubt were wounded. But it seems certain that this number included Irishmen serving in other corps. They would be easily intermingled, especially when their uniforms were similar. We know of one instance in which this happened—namely, that of MacSwiney, who was serving as a private with the Franco-Belgians. He turned up wounded at Loreto. Nicholas Furey, of Limerick, was also serving with the Franco-Belgians, and he also was wounded. Both these volunteers won decorations, and in the *Gazette* they appeared among the Irish decorations. It seems probable that Woodward and Dunne were also with the Franco-Belgians, and there were various other Irishmen as well.

The prisoners are usually stated as if they had been classified by battalions. But primarily they were separated out according to their nationalities—for the purpose of being sent home. Thus the 103 Irish included, not merely those of No. 4 Company, but any other Irishmen as well: probably all the English speakers. We know that there were at least about half-a-dozen Irishmen with the Franco-Belgians, but on the list of prisoners there are no Irish stated among the Franco-Belgians, or with anyone else. MacSwiney had, no doubt, rejoined his own fellow countrymen, and probably Woodward also and others.

The following was the report sent in to Cialdini by General Leotardi, who classified the prisoners:—"In spite of my efforts I was not able to obtain [from the C.O's] all the states (statements) of the various nations to which the prisoners belong: but.... I succeeded in compiling the state enclosed herewith, whose accuracy I cannot entirely guarantee, but which I believe to be very near the truth."—Italian War Office. U.S., Vol. 9, p. 54. It was by *nations* that he divided them.

montese at 88 dead and over 400 wounded; that is to say, about 20 percent, of those actually engaged. And if we omit the Austrians, who only came under fire at the end of the fighting, the infantry in the firing-line must have lost about 26 percent, of their number.

CHAPTER XIX.

THE FOUR IRISH COMPANIES DURING THE SIEGE OF ANCONA.[235]

WITH the siege of Ancona we come to the last episode in the war. Perugia, Spoleto, Castelfidardo, and Ancona are the principal names in this small campaign, and, by a strange turn of chance, our volunteers had had at least one company present at each engagement. Of the four companies in Ancona we have hitherto heard little because they had no mishaps to chronicle; their early risings and their sultry afternoons in that sunny hillside port were probably not very different from those of recruits in any other garrison town in Italy.

The siege has always been counted as dating from September 12, because it was from that day[236] that the state of siege was formally proclaimed in the town; but as a matter of fact the first cannon-shot was not fired until September 18, actually during the battle of Castelfidardo. This shot came from the fleet, as the result of an arrangement between Cialdini and

---

[235] For the siege of Ancona there were nine or ten veterans to be found in Ireland. Authorities: almost all those already mentioned in connection with Perugia and Castelfidardo. Be sides these there is Persano's Diary (*Diario privato politico militare*) for the action of the fleet which is very important here. On the Papal side perhaps Un Romano (*Castelfidardo e Ancona*) and De Quatrebarbes (*Souvenirs d'Ancone*) are the most necessary.

[236] *Il Piceno* of September 12th, 1860, gives de La Moricière's proclamation. The Ancona veterans were sometimes disposed to claim that the siege had lasted "about three weeks"—namely, from September 12th to the 30th.

Persano that the men-of-war should bombard Ancona so as to prevent the Papal troops under de Courten from sallying out to help de La Morcière, and it produced the desired effect.

On September 18, the evening of the battle of Castelfidardo, Persano's shells were still dropping into Ancona when de La Morcière arrived at the gates, having galloped past the Piedmontese with several officers of his staff; these few were all that remained of his army. At this moment of defeat—the first defeat of his life—he gave an extraordinary proof of endurance, both physical and mental. Although he had been very unwell when the campaign began, and since then had gone through a week of constant strain ending up with a day of battle and disaster, he showed no signs of discouragement as he rode through the gate. But then a most unfortunate event occurred. The Papal troops, having heard the distant firing all day, and seeing their General arriving, assumed that he must have fought a victorious action and their guard turned out to greet him with their bugles sounding the fanfare and the men cheering. This triumphant reception was perhaps the only one which could have shaken his self-control; he broke down for a moment as he turned and said to those next him: "Je n'ai plus d'armee," and then went straight down to his quarters, slept in a chair for half an hour, and woke up as fresh and vigorous as if nothing had occurred. That evening he held a council of the C.O's of his various units to talk over the details of the defense.

The siege of Ancona will always be remembered because it marks the final passing away of the ancient historic Papal State, and also, from a purely military point of view, because it occurred at a very interesting period in the history of war, namely that in which rifled artillery was superseding the old smooth-bores, and steamers were taking the place of the glorious old men-of-war of the days of Nelson. Rifled guns had scored their first success in the previous year at Solferino; but unfortunately for de La Moricière he possessed no rifled artillery; there were 129 guns in Ancona, but almost every single one of them was out-ranged by those of the Piedmontese, both by land and by sea. To this initial advantage possessed by the invaders, one must add the great increase of mobility and striking-power conferred on the fleet by the use of steam: also the item that, as both these factors were new in war, de La Moricière had no statistics to work on when preparing his defenses, and that, in consequence, his harbor fortifications were by no means as reliable as they would have been ten years, or even five years earlier. This was the first siege in which a fleet of steamers used rifled guns for shelling forts, and it was one of the first in which such a naval attack ever proved successful.

The story of this siege being somewhat tedious, we will deal with it as shortly as possible. The port of Ancona is enclosed by a crescent of steep hills that rise straight from the water's edge; and the houses and circle of old walls are built on the face of these hills, looking out westward[237] over the harbor—almost as easy for the ships to hit as the backing of a target In order to enclose the small harbor, two long moles had been built right out into the sea, leaving only a narrow entrance between their extremities: and across this entrance de la Moricière had stretched a powerful chain which effectually prevented any ship from coming in. On the end of his right-hand mole there stood a little tower, the Lanterna, fairly strong and capable of holding eleven guns; but, for the defense of the Mole he relied chiefly on the fire from the forts on the tops of the hills, and their fire would doubtless have proved effective if the attacking ships had been armed only with smooth-bores and therefore obliged to come to close range.

On the land side there was a zig-zag line of fortifications

---

[237] It is enclosed behind a hump of land which sticks up into the Adriatic rather like a rhinoceros' horn; this accounts for its facing westward.

from sea to sea across the peninsula, with strong forts on each flank of the central valley which leads up to the principal gates in the old wall. At its southern end—that is to say, on the right of the Papal line looking inland—stands the Citadel, a massive red brick fort on the highest point of the hills, dominating the whole scene both landwards and seawards. The Citadel was occupied by two companies, one Austrian and the other Irish.[238] Next on its left front comes the Entrenched Camp, surrounded and guarded by a foss about 20 feet deep faced with red brick-work; this was occupied by three Austrian companies and by three Irish companies, but between them they furnished two companies and reliefs to occupy the Lunette Santo Stefano, which was a strong work about 500 meters outside the walls. A Bohemian officer, Major Prossig, was in command of them all.[239]

We can picture our Irish volunteers stationed up on these forts and surrounded by guns: on the citadel were twenty-five guns, in the Entrenched Camp were thirty-two, and in the Lunette Santo Stefano were five, each with its complement of Italian or Austrian gunners. But the batteries were not up to full strength, so, when the artillery duel began, some of the Irish were ordered to assist the gunners by carrying up ammunition, and one of them told me that he was doing so under fire during several days towards the end of the siege. Some of them were hit, I think, while "shifting the shot."

These three units of fortification, the Citadel, the Entrenched Camp, and the Lunette Santo Stefano, formed the kernel, the main stronghold, so to speak, of the defense. On the opposite side of the valley (east) were two other forts, the Cappucini and the Gardetto, also important. Besides these there were three lunettes outside the walls to delay the assailants: one

---

238   It is not certain whether the Irish were all up there at first; but they were certainly all up there during the last five days of the siege.
239   De Quatrebarbes refers to him as "Prosig aussi loyal que brave."

on Monte Pulito, about 1,400 meters distant, another in front of it on Monte Pelago about 2,000 meters from the Citadel. These were not strong enough for permanent defense, but they would have to be taken before any attack could be made on the Citadel itself

Fanti's plan of operation was as follows:[240] his front was to be divided into two halves; Della Rocca with the division that had taken Perugia—originally 12,000 strong—was to be on the right, and Cialdini with the army of Castelfidardo was to be on the left; Cadorna with his central column was to be under command of Cialdini, who would thus have over 24,000 men, making a grand total, probably, of almost 34,000 men in all.[241] Fanti's first main objective was to be the Gardetto fort: but it had to be reached in three successive springs. In order to get near it, Della Rocca's men—his right wing—must first capture the outlying works on Monte Pelago, and next take Monte Pulito, neither of which posts was very strong. They would then be far enough advanced to bring up their siege guns and shell the Lunette Santo Stefano which was much stronger, and the Gardetto itself. Meanwhile his left wing under Cialdini was to keep the Citadel and the Entrenched Camp occupied by bombarding them all day at long range (from Monte Posatore); and Cialdini might also push on his own attack gradually until he could take the Lazzaretto barracks (down on the harbor) and capture the Porta Pia close by.

The fleet was to try and force the entrance to the harbor, and to keep all the forts constantly under shell-fire.

On the Papal side de La Moricière's plans were limited by his armament:[242] his guns did not range far enough to cover the outlying posts of Monte Pelago or Monte Pulito, nor to reply to Cialdini; all that he could do was to wait until the

---

240   Fanti's report.
241   They had lost some hundreds during the fighting, and several units were absent guarding prisoners.
242   De La Moricière's report.

enemy came to about 1,500 meters distance, and then concentrate all his smooth-bores on to their batteries. His garrison consisted of about 5,700 men[243], and 129 guns, which was a small number as compared with that of the Piedmontese. His walls were about 7,000 meters round; the population inside them about 25,000 souls, with 5,000 more in the defenseless suburb of Borgo Pio.

The diary of the siege may be briefly summarized:[244] September 18.—The battle of Castelfidardo raging; Admiral Persano makes his appearance outside Ancona with a line of seven battleships—six steam frigates and one sailing-frigate—and three transport steamers. Of the frigates three were screw steamers, one of these being the *Maria-Adelaide,* his flag-ship, and the other three were driven by paddles. The *San Michele* was a sailing-frigate which had sometimes to be towed in or out of action.

On this day these vessels defiled in front of the town until they reached their appointed posts, and then opened fire simultaneously on the forts above them. The Papal batteries replied, and the fight continued until evening. This first bombardment produced no great loss of life. There were five Austrians dangerously wounded, one woman and two children killed, and one civilian wounded. But it produced a great moral effect; the population retired into the cellars, and the garrison at once realized that the fleet was a far more powerful factor than had been expected; it not only out-numbered their guns but also completely outranged them; some of the ships could throw a shell 3,000 meters, whereas—according to their own accounts, at all events—the Papal artillerymen could only make effective shooting up to 2,000 meters at the most.

Anent one company of our Irish volunteers during' these

---

243   This is the figure given by the Italian translator of *Un Romain.* For many reasons it seems unlikely that de La Moricière had more than about 6,000 fighting men.
244   *vide:* Persasno's diary, Un Romano, De Quatrebares, and others.

first two days we have an interesting account in the memoirs of its captain, Count Russell of Killough, who gives the following description, beginning on September 17, the day before the battle of Castelfidardo, when the Piedmontese were only approaching the town.

> On September 17, my company (the 4th of the half-battalion of St. Patrick detached at Ancona) was ordered to the outposts in order to relieve the native [Italian] troops who had been holding them for three days. The spot was fifteen hundred yards from the town, near a brick tower situated at the point where the road to Senegaglia turns at a sharp angle northwards. From the picturesque point of view the position of this post is fairly fortunate. Sheltered by a few old trees planted in pentagon shape before the little house by the kiln, it looked out upon the open sea, and, on its western side was backed by a hill upon whose summit I placed two sentries as an observation post.
>
> In addition to my hundred and twenty men I had a section of two pieces of artillery...
>
> These precautions, one may add, seemed to be necessary owing to the danger which threatened us of being attacked at any moment.
>
> During the first night alarms innumerable kept me awake; but very fortunately they had no real cause.[245]

Though, of course, they might have had one, for Cialdini's Piedmontese lancers were expected to be in the neighborhood.

He then describes how "jumpy" his sentries were at first during the hours of darkness—a state of nerves which, he

---
245   Russell of Killough, p. 152.

remarks, was excusable: "If one puts oneself in the place of these peasants of only twenty years old, suddenly brought face to face with a real and unceasing danger, and finding themselves five hundred yards from all help, away alone in the middle of the deserted countryside—to say nothing of the burden of individual responsibility which was new to them."

Everyone knows that these long hours of night duty are a very great strain on the sentries. Thus there were constant alarms during the night, each of which caused great hilarity. Firstly he tells us that a belated pig charged into a sentry on its headlong passage home in the dark; on another occasion a swaying poplar was mistaken for a Piedmontese lance, and later a firefly close at hand was taken for a lantern further away; of course, a firefly would have been a complete novelty to an Irishman of those days.

On the 18th the Irishmen, from their post outside the town, witnessed the approach of the hostile fleet, which they hoped was Austrian, until it hoisted the Italian colors. They saw the first bombardment of Ancona. "The heavy rifled guns were throwing on to the forts of the Capucini, of Monte-Gardetto, and of Monte Marano balls weighing 130 pounds, some of which burst in the air with a sinister noise, while others passed over the town and went falling even into the Entrenched Camp, which was on the far [inland] side of the walls." Count Russell also speaks of "the long parabolas described in space by the bombs of 150 pounds weight," and he ascribes to this bombardment the fact that there was no sortie towards Castelfidardo where the battle was raging.

It was an interesting evening for him and his companions. At nine o'clock that night his two friends, François de Maistre[246] and Captain de France, arrived from the town to bring him their story of the battle of Castelfidardo, in which they had

---

246 De Maistre, and apparently de France also, were with de La Moricière's staff; de Maistre was an old friend of Russell's.

been taking part all day as de La Moricière's gallopers.

> First they spoke of the brilliant attack of Pimodan's division, followed by the offensive of the whole of Cialdini's army; then of the intrepid valor of our young compatriots,[247] most of whom were under fire for the first time, but were heroes by instinct; then of the death of Pimodan, the Bayard of the Pontifical army; then of the fine resistance of the German battalion ordered to cover the retirement; and finally they told me of the modest but useful part that my friend, G. (*sic*) D'Arcy—I had almost said my protege, for it was I who received him on the day of his first arrival at Macerata—had played at the head of his eighty Irish, during the attack on the now celebrated hay-ricks; intermingled with the Zouaves, and ready like them to fight and to die, the sons of St. Patrick had retired only when all further advance was impossible.

All this the two officers told to Russell with tears in their eyes;- and as they spoke suddenly a cry was raised, "the enemy! the enemy!" followed by several shots. Instantly every one of the 120 men sprang up and fell in, and a patrol was sent out to explore, because Russell thought that Cialdini might be taking advantage of his victory to push on an attack, in which case this outpost was full in his way.

But then a few moments later amid uproarious laughter in which even the saddened staff officers took part, a sergeant reported that the alarm was due to one of the sentries who had mistaken for Piedmontese rockets what were in reality some genuine falling stars.

---

247   Russell is including himself with the other Frenchmen with whom he is talking.

This Irish company (No. 4 of the Ancona companies, or No. 8 of the battalion) remained for two more days on outpost work. At mid-day on the 20th it was relieved and marched back into Ancona. About the other Irish companies we have no information[248] except in a letter from Lieutenant de la Hoyde. From his descriptions it is evident that they all regarded the Piedmontese invaders as immensely superior in force, and thought that they themselves would be lucky if many of them got back to Ireland again; but he seems to have been much more at ease once the bombardment began. This was his first experience of being under fire, at the age of eighteen, and it is very interesting to compare this letter with another which he wrote seven years later, before the battle of Mentana—in which engagement he distinguished himself—and to see how the home-bred, affectionate boy had developed into the calm professional soldier who had regulated his account with Providence and felt himself perfectly ready to accept the chances before him. From his letters we gather also that the Irish

---

248   The following description, however, was approved many years ago by two of the old men, and therefore may be taken to represent their view more than those by foreign writers: "Imagine the Citadel on the highest part of the cliff—a huge block of massive red-brick bastions, jutting out at strange angles to dominate the town. In it is a company of the Irish who look down all day on the houses and on the soft blue of the Adriatic below, as blue as the sea at Killiney on a fine day, now dotted with frigates and men-of-war, from whose sides come white puffs of smoke as they bombard the batteries in the port, or send a message upwards. From the land, too, it is bombarded; from the ridges of the Umbrian hills which look rather like successive lines of incoming breakers on a rough day. Upon that summit the Irish are at their post all day, in the blazing sun. During the first attacks they have leisure to watch the artillery duel beneath them, but as the siege wears on, and the enemy come closer, their own position becomes more and more dangerous; at times they are shelled from three points at once, and then the smoke goes up from the old Citadel as from a cauldron. Next to the Citadel, but on a lower level, is the Entrenched Camp, that is to say, a large expanse of short grass enclosed by a sunk fence about twenty feet high, with a heavy red-brick facing. On the Entrenched Camp are two companies of Irish lying behind banks of earth watching the shells fall about them, and their own artillery replying to the Piedmontese fire. As a general rule the hours drag on slowly under this continuous dropping of missiles, which is not accompanied by great loss of life, but, like all shell fire, is very trying to the nerves."

were greatly stirred by the eulogistic description which de La Moricière had given of their comrades at Castelfidardo:

> The natives and Swiss refused to charge... the Franco-Belgians and ours got then the order and did it like lions. One company of the Irish smashed three companies of the enemy in pieces, and then, along with the Franco-Belgians, charged the main body, twenty to one, and in the confusion the General with about 200 Light Dragoons and the Guides (about 40) charged and cut his way through the enemy.[249]

Exaggerated as this may be, it came from those who had been at Castelfidardo, and it brought many handshakes to our volunteers in Ancona.

September 19 to the 21st.—According to the Papal writers, the bombardment continued, though not from the whole fleet. Both de La Moricière and de Quatrebarbes say that their losses averaged from twenty to twenty-five men per day.[250] On the other hand, Persano speaks only of a blockade, and the Italian official history says that a blockade was "actively maintained."

September 22.—General Fanti and Admiral Persano were maturing their plans together. The army was about to appear on the scene. It had been moving towards Ancona ever since the battle of Castelfidardo five days before.

At 11 p.m. that night Persano ordered the *Governolo* to open fire at the forts from 3,000 meters distance, and she continued to shell them all night.

September 23.—The *Governolo* was relieved at 6 a.m. by the *Carlo Alberto,* which continued the bombardment until 2.30; the *Vittorio Emanuele* then came and took her place.

---

249   De la Hoyde letters.
250   *Un Romain* says that on the 20th the fleet fired shots from time to time, but only hit two gunners; and that on the 21st it had vanished; but returned on the night of the 22nd.

During this prolonged bombardment the town suffered some damage,[251] but the military losses were comparatively light. Two guns were dismounted on Monte Murano. On the other hand, the *Carlo Alberto* had been hit forty times, but there was no damage done that could not be repaired at sea, and she signaled only two men wounded.[252]

September 24.—The army began its advance. On the left Cialdini took the Scrima redoubt, but it was not seriously defended, being 1,500 meters from the town, and he began at once to establish his batteries there. On this day Count Russell's Irish company was moved up into the Entrenched Camp (next the Citadel) where de La Moricière was establishing his G.H.Q. with a view to the coming land attack. At this point Count Russell gives a rather interesting little anecdote, descriptive of life under fire, as experienced by his companies. The following is the story:—

> Madame G—[253] was the wife of a Bavarian officer who had been attached to the Regiment of St. Patrick because of his perfect knowledge of English, and also on account of the extraordinary influence that he had gained among the Irish in spite of their extremely exclusive national sentiment.[254] She was the daughter of an English nobleman and was a Protestant; and—poor young woman—had lately lost an adored child whose gentle gaiety had been her chief comfort through a period of many sorrows... so that, in spite of the gracious smile which was the char-

---

251   *Un Romain* says that 50 of the townspeople were killed and many wounded: this sounds incredible. But he adds that almost every house was damaged.
252   Persano's report. Sept. 23—Uff. Stor. I. p. 289.
253   Baroness Guttemberg, whose husband was in command of No. 5 company of the Irish battalion (No. 1 company of those in Ancona.)
254   He had convoyed them from Ireland to Italy.

acteristic expression of her well-bred features, it was easy to see that life weighed heavily upon her. Several times she had started upon religious discussions with me, and in these conversations, during which she showed a well-informed mind and excellent powers of conversation, I had been led to think that her religion did not satisfy either her generous aspirations or the anxiety of an intellect which hungered after faith. When her husband, to whom she was sincerely attached, had gone up into the Entrenched Camp with the Irish company under his command, she had insisted, even against his strongest, appeals, on accompanying him, in spite of the fatigue involved, the danger of each succeeding hour, and the horror of constantly seeing wounded men being carried to the ambulance.

One day, when the bombardment was hotter than usual, and the "death-bottles" [bouteilles de mort], as our men named the conical bullets, were raining everywhere upon the main traverse into which they finally vanished after finishing their furious flight through the air, what was my surprise, on arriving at Major Prossig's quarters for orders, to find Madam G— sitting on a thick canvas stool, and talking to the officers as calmly as if she had been in her drawing room, impervious apparently to the deafening noise all round her.

The C.O's hut, which stood higher than the others and was surmounted by a little red flag, afforded a specially suitable point of aim for the enemy's artillery; and consequently the shells were humming by without a moment's rest or truce; there was, in fact, on the side of Monte Scrima, a battery of four guns whose fire was exclusively directed against us—a favor with which we should have been very ready to dispense. Under the fragile shelter of the planks we had become able, to distinguish with fair accuracy, by the direction or intensity of their sound amid the general uproar, those shots which came from this terrible battery, and then, instinctively, there would be a pause in the conversation until the strident whistling caused by the projectile had ceased. The only one whose talk did not cease was the young woman; she seemed to chaff us when we bowed to the projectile, as is customary.

"Why, gentlemen," she would say, accompanying her words with an ironical curl of the lip, "do you believe that death can take you by surprise without its having been foreseen and ordered by God? And in that case even your lowest bow will not prevent the providential messenger from hitting you."

This fatalistic reasoning, as may be imagined, did not reassure Captain G— as to the safety of his wife, whose stoicism reduced him to despair, especially as the enemy's fire, which had been uncertain at first, was every moment becoming more and more accurate. By every means that tenderness could suggest, he begged her, to retire. But in vain. She insisted on remaining—in the hope, perhaps of death's coming to take her away to the child that she mourned.

This condition of affairs could not last indefinitely. The poor mother had just contemptuously rejected our respectful remonstrance when a hollow bullet came through the sloping roof of the hut, passed close by the head of the young woman, happily without touching her, and buried itself in the ground a little way in front; it then burst in a score of pieces, covering us with earth, smoke, and shapeless debris. We dragged Madame G— out of that place of death by main force, and her husband got her a casemate in the citadel, after making her promise, not without some trouble, that she would stay inside during the bombardment.[255]

September 25.—On the right of the army Della Rocca's men began their advance and took Altavilla hill, defended only by a few men, and built an epaulement for a battery, with a view to their attack on the Lunette; but they failed to silence the Austrian gunners at Monte Pelago, and were obliged to postpone their attack.

On the left Cialdini's men, near the Scrima redoubt, began shelling the Citadel and the Entrenched Camp, which were manned, as we have said, by the Austrians and the Irish, from a distance of 2,200 meters; and under this direct and heavy fire from the land side, our volunteers must have had an unpleasant time. It is thus described by de Quatrebarbes:

> Their hollow bullets and shells were intermingled with those of the fleet. They were principally directed against the Citadel, which they covered with a rain of fire.

But de La Moricière had long laid all his plans for taking advantage of the Piedmontese as soon as they came within

---

[255] Count Russell of Killough. *Dix annies*, etc., p. 169.

range. He at once concentrated a large number of guns on to the Scrima, and claims that in two or three hours he had completely silenced the Piedmontese battery there.[256]

On this day, therefore, de La Moricière had succeeded in checking the attack in both sectors, but the fleet had done him some damage, dismounting several of the Papal guns and inflicting losses on both the soldiers and the civilians.

September 26.—This was the day on which the most serious fighting took place and it marks a considerable advance for the attacking force, although that had been partly foreseen.

The day began, strangely enough, by a Papal attack delivered before dawn—an attack of which the cause was rather curious. In front of Monte Pelago lay the small village of Pietra della Croce, one end of which was occupied by the Piedmontese troops and the other by a company of Austrians. But with the Austrians were two companies of the Papal Composite Battalion sent out from Ancona under command of a gallant Swiss officer, Captain Castellaz. Castellaz had been greatly hurt by some doubts expressed by de La Moricière as to the reliability of his men, and he had obtained leave to attack the far end of the village and to try a coup-de-main on the hill of Altavilla beyond it, in order to prove their loyalty by some successful feat of arms. Before daybreak, therefore, he started with two companies, cleared the village[257] of the enemy, and threw himself on to the Piedmontese outposts. But the two Bolognese companies there stood firm, whereas most of his own men gave way, so that he narrowly escaped capture and was obliged to retire to Monte Pelago, and thence to Ancona.

The results of this retirement were considerable. The two captains of the Bolognese outpost companies happened to

---

256 The Italian official history (p. 443) says that owing to the Piedmontese guns not being kept cool with water, according to orders, three of them burst and two others became unserviceable.

257 The Italian official history says that, as a matter of fact, they had no troops there.

be aware that their commanding officers were proposing to take the village of Pietra della Croce, and when they saw the composite Papalini turn tail, they were struck with a sudden inspiration.

In a very short time they had advanced and "rushed" the village, surprising and driving out the Austrian company which alone remained to defend it. This achievement, so rare in the history of outposts, seems to have inspired them still further. Following closely upon the heels of their retiring enemy, they actually got right up Monte Pelago, and it was only when they were quite close to the redoubt itself that the Austrians inside it were able to check them, and even then, though repulsed, the Bolognese officers only retired for a short distance and sent word to their C.O. that Monte Pelago could be taken if reinforcements were brought up.

Their C.O., General Pinelli, was at that moment actually engaged in considering this very problem, and when the message arrived he at once launched his whole brigade to the attack, that is to say, about three thousand men in all. Now, the Austrians in the redoubt had orders to retire if they were attacked by preponderating forces and not to lose their guns, so that they were perfectly justified in withdrawing when they saw this torrent coming at them, especially as their line of retreat was already threatened. But they were loth to go, and in the end they stayed rather too long, and then became a little unsteady and fired high, and finally, after making some slight defense of their guns with their bayonets, they were compelled to spike them and leave them behind. Thus by the initiative of two junior officers the Piedmontese had captured the Monte Pelago redoubt with very slight loss and in it five Austrian guns. They planted on it the colors of the 39th Regiment of the Bolognese Brigade, amid the cheers of the men of the 5th Army Corps from all the heights around.

But even there the two companies of the outpost were not

satisfied. With the agreement of the General they at once led the way in a fresh rush towards. Monte Pulito, six hundred yards further forward, followed as rapidly as possible by Pinelli and his brigade, with the exception of two battalions left to occupy Monte Pelago. This rush was equally successful. The three companies of Austrians defending Monte Pulito had evidently taken warning by the fate of their comrades, because they retired before the Piedmontese arrived, and consequently succeeded in getting their guns away with them. The colors of the 40th Battalion, Bolognese Brigade, were planted over Monte Pulito.

Having secured Monte Pulito, the enthusiastic column of assault pushed forward once again, led this time by a Bersaglieri captain, up past the cemetery with the intention of rushing the Santo Stefano lunette only 500 meters from the Citadel. But in this instance they were tempting Providence; the Santo Stefano lunette was stronger than the other two, and was within easy range of the Papal batteries in the forts; and de La Moricière was directing its defense in person. The following is the description which he gives of this episode, and it is of especial interest from our point of view, because the Lunette Santo Stefano was defended, as already stated, by one of our Irish companies together with one of the Austrians,[258] and was under the command of Major Prossig, C.O. of the Irish half-battalion in Ancona.

In obedience to de La Moricière's orders, the garrison

---

258 During' the whole siege the Irish half-battalion and Austrians may be said to have fought side by side. Their commanding officer was Major Prossig who was in charge of the Citadel, Entrenched Camp and Lunette Santo Stefano—Major Prossig, a Bohemian by birth, had been an officer in the best of the Austrian Papal battalions, officially known as the 2nd Bersaglieri; the one that fought at Castelfidardo. *V. De La Moricière's Report,* pp. 42 and 43. And *Un Romain* in *Castelfidardo e Ancona* (Italian translation) speaks of an Irish and an *Austrian* company charging side by side during the sortie (p. 184). And speaking of the night of Sept. 29th he says: "General Kanzler was awaiting the attack, with the Irish and an Austrian half battalion posted in the Entrenched Camp." Moreover, the Irish veterans spoke to me of Austrians being with them.

allowed the Piedmontese to advance right up to the edge of the lunette, and did not fire until some of the leading assailants had actually reached the foot of the escarpment.

"The enemy's skirmishers," he says, "advanced openly, and the bravest of them descended to the foot of the escarpment. Then a terrible fire flamed out on them from all sides, from front, from flank, and from rear, and they were obliged to retire in disorder. They made a brave attempt to rally behind the hedges and houses, but the bullets and shells soon drove them out from there, and they were not able to re-form until they got back behind the redoubts that they had taken some hours earlier. A few mounted officers whom I had seen lead this assault with a want of foresight only equaled by their daring, directed the retirement from this attack, which had cost them a great many casualties."

A sortie was then ordered, and the Irish company dashed out with fixed bayonets and charged after the enemy, cheering; but, by the time they were able to get into the open and form line, most of the Piedmontese were already out of reach, retiring towards the cemetery.[259]

---

[259] This sortie is variously described according to the sympathies of the writer. The O'Clery says it was ordered to cover the retirement of the Austrians from Monte Pelago—and it is quite possible that the rear of their column may still have been on its way. He says: "The movement being covered by a dashing charge made by two companies of the Irish who took fifteen prisoners." Un Romain gives a similar account (the two companies "lowered their bayonets and two or three times attacked the enemy, who had to retire before their bold attack"), but the translator of Un Romain (an Italian Papal officer) denies that they took any prisoners; de La Moricière makes no mention of the sortie, but this is not surprising as he was in the Citadel, and the order was doubtless given by Major Prossig, or by whatever officer was in command on the spot. But it is mentioned, though as briefly as possible, in the Piedmontese reports and books, especially by Carandini, a strongly Piedmontese writer who wrote a life of Fanti; after dealing very sparingly with the Piedmontese repulse, he says that it was followed by a sortie. "Soon afterwards a strong sortie threw itself on the front of the four battalions of Bersaglieri. How it was received by them one can easily imagine. After the repulse of the sortie, fire was resumed from the enemy's forts."

But of the six or seven Irish veterans of Ancona, there were only two who had taken part in this sortie, and apparently in different companies. One of them, Patrick Callaghan by name, said to me: "when we shouted, the Piedmontese were

It is said, however, that they made a few prisoners.

On the Piedmontese right, therefore, the net result of that morning's work was that Della Rocca's men had advanced about 2,000 meters, capturing the outlying defenses of the Papal lines. But they had also been compelled to realize that the actual fortifications of the town, covered by the guns of the forts, were not to be taken without artillery preparation, so they began at once to establish batteries, including some of their siege guns, on the site of the conquered positions.

Meanwhile, on the Piedmontese left, Cialdini's men were pushing forward into the Borgo Pio, a large suburb whose houses were packed close up to the walls of the town, and therefore very difficult to defend. Cialdini aimed also at capturing the Lazzaretto barracks, situated on a peninsula running out into the harbor quite near to the Porta Pia gate. Once established in the Lazzaretto barracks, he could from there bring fire to bear on the defenders of the gate.

During this day his batteries on Mt. Posatore and near the Scrima redoubt brought heavy fire to bear on the Citadel and the Entrenched Camp, and we have a vivid description of the behavior of our Irish volunteers under shell-fire. An old French officer, Count de Quatrebarbes, Governor of Ancona, speaks of them as follows:—

> Some days earlier the General had withdrawn the half-battalion of St. Patrick from the

---

running back," meaning that by the time their charge began the enemy were already in full retirement. This seems to me to be the true version.

One of the other old men had been either in the Lunette Santo Stefano or, more probably, in the Entrenched Camp: he (Michael Fallon by name) wrote to me that he remembered that " when the guns opened fire the enemy broke away from each other [extended ?] and we were ordered out to follow them, with fixed bayonets; when we were some time [gone?] they opened fire again, but not for long," and he subsequently confirmed this by word of mouth. The O'Clery speaks of two Irish companies being engaged in it. And he is probably right as to their making some prisoners; they would have been almost sure to capture a few of the retiring enemy; some of the wounded men, if not others as well.

> Lanterna and the Mole.²⁶⁰ Feeling assured of the courage and devotion of these brave sons of Ireland, he had placed them in the posts of honor, namely, the Citadel, the Entrenched Camp, and the Lunette Santo Stefano, to face the principal attacks of the enemy. Nothing was finer than their bearing when the volleys of bullets from the rifled guns whistled over their heads; they greeted each volley with a cheer for Pius IX, and sang in chorus the old ballads of their mountains, and shouted challenges at the Piedmontese; and it was all that their officers could do, our brave captains Guttemberg, Russell, and O'Mahony, to prevent their coming out from behind cover and climbing the cavalier of the Entrenched Camp to jeer at the Piedmontese and applaud any lucky hits of our gunners. Their courage never lessened during the siege. And Major Prossig, the officer in command of the camp, proved himself in every way worthy of commanding them.

This account is valuable as coming from an outside source, and it is confirmed by that of Russell of Killough, who thus describes the life of his men:

> During the first days we had no shelter; no tents or huts: hardly any blankets; and to add to our discomforts, the earth, which had been dug not long before in order to construct the works

---

260 It seems that as long as the attack had been only from the fleet he had left the Irish at the harbor mouth to repel any attempt at landing but when the land-attack came closer he brought them up into the forts to repel it, and left the Lanterna and the Mole entirely to gunners. Count de Quatre barbes refers to this fact in a complimentary manner, as will be seen in the words that we quote from his book, *Souvenirs d'Ancone*.

ordered by the General, was turned into thick mud by the first rain that fell. It was on this bed that the men had to sleep, with no shelter but the vault of heaven; and yet, contrary to what I had feared from previous experience, there was no grumbling. Throughout the whole night they remained on this soaking ground, exposed to a driving rain, without allowing themselves even the slightest murmur—so greatly had the sight of the enemy excited their ardor and developed their military virtues.

Strange! The falling of a few bombs had sufficed to turn these intractable peasants of yesterday into sober, patient, war-hardened soldiers capable of any sacrifice.

He then describes how in the evening they used to sit round the fire and sing, at the risk of getting a shell among them.

During the day's fighting the Piedmontese had lost 15 men killed and 11 officers and 83 men wounded; the Papal troops, about 40 killed and 150 wounded; of the Irish, the company in the Lunette Santo Stefano seems to have had seven or eight casualties, and the company in the Citadel four or five. The following short description of the scene, written years ago, was approved by three or four of the old men who supplied the information:—

> In the above picture (de Quatrebarbes') we see them singing and larking as the shells fall about them; but presently the news goes round that someone is hit. That shell burst in the Citadel. Private Gorman is already on the casualty list with a bullet through the shoulder, and Nevin with part of his foot blown off; now it is a boy called O'Beirne who has fallen with a shattered

leg. Three or four of them carry him off to the hospital, and—a proof of how careful an officer should be not to let the men see their wounded if he can help it—one of these carriers has told me that although they went to the door full of courage, the sight of the doctor at work took all the steadiness out of them and they came back feeling sick and nervous. But young O'Beirne himself remains full of pluck. When he is told that his leg is to be amputated—without chloroform, of course-yall he asks for is his gun to grasp during the operation, and adds, "If I die, anyhow it'll be like a soldier."

I think he died in Paris on the way home. And another wounded man, a sergeant named Skehan, died on arriving in Cork.[261] Very likely there were other deaths, but no casualty-list has been kept.

September 27 was a day of comparative respite; Della Rocca was busy getting up his siege guns to Monte Pelago, and Cialdini in establishing himself in the Borgo Pio, and planning a coup-de-main on the Lazzaretto.

The fleet shelled the forts as usual, but after a few hours their bombardment was stopped by a tornado of rain; and the Irish in No. 8 Bastion seized upon the moment to build a new traverse. They spent the rest of the day toiling with great enthusiasm in the muddy ground, and before evening the work was complete.

September 28.—At about one o'clock in the early morning Cialdini's Bersaglieri, with a few engineers, surprised and captured the Lazzaretto. This was an adventurous exploit. About a half-company of them went out in a boat and climbed

---

261   For this information I am indebted to Mr. Cremin, now living in Cork. He remembers Skehan's funeral.

into the place through an embrasure. They were rewarded by the good fortune which is said to favor the brave, for, after the first fire, the Papal troops made no resistance, but either surrendered or escaped by swimming. By now most of them were very greatly demoralized, by no means inclined to risk their lives for the sake of prolonging a defense which could not, in the end, prove successful: it was said that the Austrian and Irish units alone showed no signs of discouragement[262] and that some of the others were completely demoralized.

The assailants were now on the verge of their great attack. Fanti hoped to have his siege guns in position by the night of the 28th, and to begin then, or on the following morning, a heavy bombardment of the Citadel, the Entrenched Camp, the Lunette Santo Stefano, and the Gardetto. In fact the Irish companies were in for a bad time, and so were various other units. Cialdini in the meanwhile was to attack the Porta Pia; and that afternoon the fleet was to attack the Lanterna.

As matters turned out the latter event brought about the surrender of the place sooner than had been expected.

Towards one o'clock that day three of the frigates received the order to move forward, and about an hour later, in spite of the stiff breeze against them, they were anchored at a range of no more than five or six hundred meters from the Lanterna. The sound of their intensified fire at once conveyed the news that a serious attack was in progress, and from every height all over the field of operations both men and officers strained their eyes towards the sea, watching the contest between the "lighthouse" and the three frigates.

It was not long before the great superiority of the naval guns began to assert itself, especially on the floating batteries designed to protect the entrance. One after another these

---

[262] "Discouragement was especially prevalent in the ranks of the Foreign or so-called Swiss battalions and in the Italian battalions, whereas the Irish and the Austrians showed themselves full of keenness and ready to carry out their duty to the end."—*Un Romain*, p. 198.

platforms were sunk, until at last nothing of them remained to be seen except the tops of their masts on which the white and yellow Papal colors were still flying. Thus by three o'clock the *Carlo Alberto* was able to steam up to within 200 meters of the Mole and all three vessels concentrated their hundred and thirteen guns upon the Lanterna. In the barbette of the Lanterna there were only three guns, and down in the casemate eight, but so arranged that not more than three of them could face towards any single front. By now the Papal batteries on the hills were joining in the defense, but apparently without producing any great result. For an hour or more the plight of the Papal gunners (mostly Italians) must have been terrible; nevertheless, under a young Austrian officer, Lieutenant Westminthal, they continued to resist with extraordinary courage until most of the embrasures had been blown out of shape and many of the men were dead. By five o'clock all the guns of the barbette had been dismounted, so Lieutenant Westminthal and his junior subaltern, a young Italian named Delle Piane, led the men down into the casemate below. Here the fight continued for a while until at last the coup-de-grace was given by the *Vittorio Emanuele,* which steamed boldly up to within pistol shot of the Mole and with one broadside laid the casemate in ruins. Westminthal was dead among the masonry, and the few remaining men raised the white flag and filed out of the fort. But even then the fight was not entirely over. Captain Castellaz, who was present, at once ran up the steps to the top of the lighthouse, tore down the white flag, and hurled it with all his strength at the frigate, which was swaying on the waves only about thirty yards distant, and hauled up the Papal colors once again. It was a fine act of defiance, but useless; for with the very next broadside one of the Piedmontese shells fired the magazine, and a few seconds later there followed a terrific flash and explosion, and the tower was hidden in a rising column of smoke; and when the debris had fallen and the smoke cleared away it was seen that the fort was only a heap of ruins. Of the Austrian and Italian gunners, about 150 in number at the start, only 25 remained alive—the 25 who had filed out by order of their officers just before the end. With them were Castellaz and Delie Piane.

The whole of this last heroic episode was witnessed with profound interest by the Irish companies up in the Citadel and Entrenched Camp, but, of course, they could do nothing to help their comrades below.

Very soon afterwards the white flag was waving over the Citadel and Ancona had surrendered. De La Moricière had thought of continuing the defense in the Entrenched Camp and the Citadel,[263] but felt that it was useless to do so; not only had the Lanterna been destroyed, but the chain which he had stretched across the entrance was now at the bottom of the sea, because the end of the wall to which it had been fixed had been blown to pieces, and in consequence there was nothing to prevent the fleet from steaming into the harbor, shelling any of the defenses at short range and landing men if they so desired. In any case he had made his appeal to the Catholic powers of Europe, and the Catholic powers remained silent.

This last scene dealing with the defense of the Lanterna does not in reality concern our subject, but we have described it at some length because it sheds a ray of glory over the end of the ill-starred Papal army, about which we have been compelled to record more than one unfortunate episode.

---

263   De Quatrebarbes, p. 233.

CHAPTER XX.
THE END.[264]

WITH the Lanterna lying in ruins and the white flag; flying on the Citadel of Ancona we come to the end of the Campaign, of the Papal army, and of the Papal State. The disheartened troops, now prisoners, were sent marching off under Piedmontese escort, each man to his destination, according to his nationality. And in many places as they passed, the people rose and cursed them.

At this moment of profound discouragement there is just one tiny episode worth recalling. When all was over and the column of prisoners was marching out of the town unarmed, between lines of Piedmontese, some of the Irish came to their officer, Count Russell of Killough just as they were emerging from the suburb, and said to him that they could not leave the town without bidding it good-bye. He agreed, and a moment later, in response to his call, every man in the company turned round waving his cap to the forts. and cheering: "Pius IX forever!" The Piedmontese escort was startled by this demonstration and uncertain how to take it; but the officer in charge

---

264   My principal source of information in this chapter is the late Mr. Crean, who knew something of the subsequent history of many of his brother-officers. Other veterans were also able to help, notably Mr. Bergin and his son.

Of books, the most instructive is the *Irish Brigade* by Captain Conyngham. But Count Russell of Killough, and other writers. of memoirs, also provide items of interest.

perhaps. expressed rather more than he actually realized when. he observed: "Let them do it; they are Irish."

The remainder of the story can be told in a few paragraphs. The volunteers were marched across Italy, and were all finally assembled at Genoa, where they remained for several weeks until they were discovered by Dr. Whyte,[265] who undertook a great deal of work for the organizers in Ireland at this time. At first there seemed to be little prospect of their repatriation, but Mr. A. M. Sullivan took up the matter, and through the generosity of a wealthy Irish Catholic was able to hire a vessel and bring them home to Ireland. None of these boys had been away for more than about four and a half months, and some of them for much less, but they had learnt a very great deal in that short time, and it was perhaps due to their experience in this small campaign that many of them became officers three or four years later on one side or the other during the gigantic and fearful struggle known as the American Civil War; and won glory for Ireland in the old 69th Regiment and in other corps.

Of those who joined the Northern army, quite a large number earned distinction. Captain Coppinger, wounded at Spoleto, rose to be one of the best-known generals in the standing army of the United States, and died as he had lived, a patriotic Irishman. Keiley, a lieutenant in Ancona, commanded a cavalry brigade during the American Civil War, and subsequently obtained a captaincy in the Regular Army in recognition of his services. Keogh, another of the Ancona subalterns, rose to be a colonel on the Staff of General Stoneman. O'Keeffe (Ancona) also joined the cavalry, and was a lieutenant-colonel on General Sheridan's staff when he was killed. Luther, of Perugia fame, served as a captain until he succumbed to the hardships of the campaign, and, as in Italy, won a reputation for exceptional courage. Stafford (Spoleto) served as a captain,

---

265 His name is sometimes spelt White.

and Cronin (Spoleto) as a lieutenant. O'Connell (Ancona) was killed while "bravely leading his company at Spottsylvania." D'Arcy, who "led the men at Castelfidardo," rose to be a captain in the Papal Zouaves, and was on the staff of the unfortunate Emperor Maximilian in Mexico until the disastrous ending of that expedition. De la Hoyde, a 2nd Lieutenant in Ancona, enlisted in the Papal Zouaves, obtained a commission at the end of two years, was promoted captain in 1867 for gallant conduct at Mentana, and commanded a company during the defense of Rome in 1870, after which, of course, his career came to an untimely end. Crean, wounded while defending the gate at Spoleto, and rewarded by being made Cavaliere of the Ordine Piano, retired into civilian life and became a barrister of standing in Ireland, but the fighting tradition was carried on by his son, who won a V.C. during the Boer War. So much for the officers. Of the N.C.O's, Mulhall, sergeant-major at Spoleto, served throughout the American Civil War, distinguished himself at an engagement called Skinner's Farm, and ended up a captain. Gleeson, sergeant-major or sergeant at Ancona, rose, I believe, to be colonel. Clooney, the private who defended the house in Perugia, was a captain on the day when he fell, gallantly leading his company "where the dead lay thick at Antietam"; for over a year during that war - he had maintained a splendid record, and even in Meagher's Brigade was remembered as "the bravest of the brave." But these men can only be a few, perhaps a very few out of the numbers whose doings might be recorded if we knew the names of the rank and file as well as the officers, or the history of those who fought in the Confederate army.

 The discomforts endured by the Irish wounded proved the wisdom of those who had refused to report themselves as such, so long as they were capable of marching. They seem to have been well enough treated by the Piedmontese, but they were lost to their friends and relatives. It was only through the zeal of Dr. Whyte that they were discovered in the various hospitals in Italy, without money, and many of them short of clothing. After several months of difficulty they were finally brought home to Ireland.

The next fight of the volunteers consisted in repelling the poisonous campaign of slanders hurled against them and against the Pope by the newspapers. Of this campaign the less said the better. It has no importance from the historical point of view.

But it produced one good result. The people of Ireland were determined to show everyone that they did not believe a word of these accusations. From the day when the volunteers landed in Cork amid cheering crowds until the morning when they finally parted from one another in Dublin to return to their homes, everywhere they were received with supreme enthusiasm. It was felt that they had made sacrifices not only for the Papal cause, but also for the cause of nationality in Ireland.

Major O'Reilly returned to his home, Knock Abbey, in County Louth, and to his patriotic work. In 1862 he was elected M.P. for Longford, and soon became a popular figure in the House of Commons.

Here we come to the end of this Memoir of the Irish expedition to defend the Pope in 1860—for that is probably the most suitable title for it. To call it a history of the Battalion of St. Patrick would suggest rather a wrong idea: it would give the impression of a regimental history, whereas in reality the battalion never reached the stage of being complete. When war broke out a large number of the men had not been in possession of their muskets and bayonets long enough to learn how to use them; hardly any of them can have had their full kit; and a few, I believe, of those at Spoleto actually fought in the old civilian clothes in which they had travelled out six or eight weeks before. The surprising point about it was not the degree of success that the battalion achieved, but the fact that it achieved anything at all.

On one point one cannot insist too often, namely, that we all see now that in 1860 the Papal State was no longer a possibility, and that the forming of an Italian nation was a true and glorious enterprise worthy of the brave men who gave their lives for it And certainly many of the staunchest supporters of the Papacy nowadays go even further. They believe that the loss of the Papal State has proved a blessing to the spiritual power, even as the disestablishment of any Church may prove a benefit to its spiritual power. But these truths were not realized in 1860.

Italy is now a great nation, and she can afford to look back with a smile at this, her first campaign, and also to realize that the men who fought against her, though acting from mistaken motives, were not plunderers, as Cialdini said, nor even mercenaries, but volunteers, offering their lives for a cause in which they genuinely believed.

The End.

APPENDICES.

APPENDIX A.

CARDINAL DE MÉRODE'S FAREWELL ORDER TO THE BATTALLION OF ST. PATRICK.

This rather crude translation I copied in a farmhouse near Sligo, in which it had been treasured for fifty years. There are several other translations of the original, but this is the version preserved by the veterans.

Cardinal de Mérode's farewell order to the Battalion of St. Patrick.
General order of the Minister of Arms, of October 6th, 1860.

At the moment in which, in consequence of the present sad state of affairs, the brave soldiers of the Battalion of St. Patrick who had hastened hither for the defense of the State of the Holy Church are about to leave the Pontifical army, the undersigned Minister of Arms experiences the liveliest satisfaction in being able to express to the soldiers his entire satisfaction, and in bestowing on them the highest praise for their conduct.

Nothing more could be expected from them. The Battalion of St. Patrick at Spoleto, at Perugia, at Castelfidardo, and in Ancona has shown the power of faith united to the sentiment of honor in the treacherous and unequal contest in which a small number of brave soldiers resisted to the last an entire army of sacrilegious invaders. May this recollection never perish from their hearts. God who defends His Church will bless what they have done. It is not Irishmen who require to be reminded that we must suffer and persevere in the good fight.

XAVIER DE MÉRODE,
Minister of Arms

APPENDIX B.

DECORATIONS WON BY THE BATTALION OF ST. PATRICK DURING THE CAMPAIGN OF 1860.

The Papal government certainly distributed its decorations with a generous hand, but not so lavish as that of the Piedmontese: no less than 700 decorations were awarded by the Turin authorities for the campaigns of 1860, in which the army had no very difficult task to perform. But they had political ends in view; they wished to popularize their campaigns, and to make them seem even more glorious and more wonderful than that of Garibaldi; and it was the first adventure of the new nation, so they were doubtless anxious that everyone should be happy.

The Battalion of St. Patrick did well in the matter of decorations, and certainly received more than any other Papal battalion, except probably the Franco-Beiges; but, after all, it had a far better claim than most of the others; and one can hardly be surprised if the Papal government showed pleasure in rewarding the men who in all four of the principal engagements had been found among the faithful few.

Regio Archivio di Stato, Rome.
Archivio del Ministro dell' Armi, Busta 1169.
Ordini del Giorno dal 10 Agosto al 31 Dicembre, 1860.

Order of the Minister of Arms, November 22nd, 1860.

The undersigned pro-Minister of Arms desires to inform the Army that His Holiness has been pleased to grant the following decorations to the various soldiers of the Pontifical Army in reward for courage, faithfulness and zeal shown during the late crises .... and trusts that these signs of His sovereign clemency will inspire every man in the accomplishment of his duty.

Battalion of St. Patrick.

    Major O'Reilly (Myles), to be Commendatore of the Ordine Piano (Order of Pius).

    Captain O'Mahony (Timothy), to be Cavaliere (Knight) of the Ordine Piano.

    Captain Russell (Frank), to be Cavaliere of the Ordine Piano

    Captain Coppinger, to be Cavaliere of the Ordine Piano.

    Captain Blackney, to be Cavaliere of the Ordine-Piano.

    Captain Sweeney (serving with the Franco-Belgians at Castelfidardo), to be Cavaliere of the Ordine di S. Gregorio (Order of St. Gregory).

    Lieutenant Keiley (Daniel), to be Cavaliere of the Ordine di S. Gregorio.

    2nd Lieutenant Darcy (James), to be Cavaliere of the Ordine di S. Gregorio.

    2nd Lieutenant Stafford (William), to be Cavaliere of the Ordine di S. Gregorio.

    2nd Lieutenant Crean (William),[266] to be Cavaliere of the Ordine Piano.

    2nd Lieutenant Lynch, to be Cavaliere of the Ordine di S. Gregorio.

    2nd Lieutenant Cronin, to be Cavaliere of the Ordine di S. Gregorio.

---

266   A mistake for Michael.

Appendices

2nd Lieutenant Lloyd,[267] to be Cavaliere of the Ordine di S. Silvestro (Order of St. Sylvester).

Sergeant-Major Mulhall (Thom. Dillon), Cavaliere. of the Ordine di S. Silvestro.

Sergeant Major Deadey (William), to be Cavaliere of the Ordine di S. Silvestro.

Sergeant O'Neill (David), to be Cavaliere of the Ordine di S. Silvestro.

Sergeant Fitzpatrick (Richard), to be Cavaliere of the Ordine di S. Silvestro.

Sergeant Synan (William), to be Cavaliere of the Ordine di S. Silvestro.

Corporal MacDermot (James), to be Cavaliere of the Ordine di S. Silvestro.

Corporal Ward (Michael), to be Cavaliere of the Ordine di S. Silvestro.

Private Busteed (Richard), to be Cavaliere of the Ordine di S. Silvestro.

Private Walker (William John), to be Cavaliere of the Ordine di S. Silvestro.

Private Beirne (John), to be Cavaliere of the Ordine di S. Silvestro.

Private Murphy (Michael), to be Cavaliere of the Ordine di S. Silvestro.

Private Lions (James), to be Cavaliere of the Ordine di S. Silvestro.

Private Furey (a Limerick man, but serving in the Franco-Belgian battalion at Castelfidardo), to be Cavaliere of the Ordine di S. Silvestro.

The services for which the officers were rewarded have already been described in the text, or in the list of their names;

---

267   Thia is evidently Sergeant Lloyd (Spoleto) promoted to be a Second Lieutenant; but given the Order of St. Silvester which is the Order for N.C.O's and men; and not the Order of St. Gregory or of Pius. which are for officers.

but about the N.C.O's it may be as well to give a few words of explanation.

Lloyd, Mulhall, Deadey, O'Neill and Fitzpatrick are the N.C.O's of whom O'Reilly speaks in his diary, and whose names he mentioned in his official report. Synan, of course, is Corporal Synan of Allman's party (Perugia), but now promoted to be a Sergeant. McDermot is no doubt the "young MacDermot" mentioned by O'Reilly as being with the 23 Franco-Beiges, in Spoleto; incidentally, it may be added, that several of the Franco-Belges received rewards on O'Reilly's recommendation. Ward, Busteed and Walker are the Spoleto men mentioned in O'Reilly's report. Beirne is impossible to identify. Murphy is probably the man of that name in Allman's party, and Lions is. not the man who distinguished himself with Diamond in the Citadel at Perugia, but the Lions in Spoleto.

The Furey mentioned, at the end of the list is no doubt Nicholas Furey of Limerick, an Irishman, but one of those who joined foreign units. Furey was in the Franco-Belgian Tirailleurs; fought at Castelfidardo, and lost a leg, amputated at the thigh; and spent sad years afterwards in Rome, trying to live on a very small pension; he had made over what little property he possessed to his sister.

The above-named were the first batch of Irishmen who won decorations; but within the next few months the following lists were issued, adding to the number.

The Reference is:
Regio Archivio di Stato Roma.
Archivio del Ministero delle Armi, Busta 1169.
Ordini del Giorno del 3 Gennajo al 30 Dicembre, 1861.

> January 17, 1861. O'Connell (Michael), to be Cavaliere (Knight) of the Ordine Piano (Order of Pius).

By that date he was an officer in the Papal Zouaves; but he had been through the siege of Ancona in the Irish Battalion.

Appendices

And on June 9th, 1861, a further list was issued; and one is interested to note that it contains the name of Peter Diamond, who had earned praise during the attack on Perugia. The list was found in the Archives but for knowledge of the services rendered by these men I am indebted to an article by Mr. O'Donoghue in the Dublin *Standard,* of May 11th, 1929; he quotes from the *Morning News* in the year 1861.

> Private William O'Brien received the Ordine Piano (Order of Pius IX) a cross rarely granted to privates: he had been severely wounded in the leg during the defense of the burning Upper Farm at Castelfidardo.
>
> Sergeant-Major George Davis was awarded the Cross of St Sylvester for distinguished services at Perugia.
>
> Sergeant John Buckley was awarded the Cross of St. Sylvester for working a gun at Ancona.
>
> Corporal Michael Stack received the Cross of St. Sylvester for his services at Spoleto.
>
> Private Peter Diamond received the Cross of St. Sylvester for his services in Perugia.
>
> Private Patrick Leahy received the Cross of St. Sylvester for his services in Spoleto.
>
> Private Michael Walsh received the Cross of St. Sylvester for his services in Spoleto.

To another list of decorations also Mr. O'Donoghue's articles have called attention (*V.* the Dublin *Standard,* May 4th, 1929)—namely, the supplementary roll published on October 21st, 1861.

About this list one has to add that there appear to be no traces of it among the gazettes in the Italian State Archives, although it is there that the other Papal documents on this period are kept. It might have evaded the notice of the present writer, but it would hardly have escaped that of the Marquis MacSwiney as well. Possibly some reference to it may be found in the Vatican Archives when they are fully thrown open.

At the time of his first report Major O'Reilly had not been in a position to discover the names of all whose work deserved recognition—least of all those at Perugia and Castelfidardo. On October 21st, 1861, therefore, a supplementary list was published, the first name being that of Lieut. Luther, awarded the Cross of St. Gregory for his coolness under fire at Perugia.

The list of N.C.O's and privates who were awarded the Cross of St. Sylvester is as follows:—

Sergeant Thomas Lyons (Perugia), Privates James Doran (do.), Tames Ryan (do.), Philip Kirwan (do.), Thomas Kirwan (do.), Michael Smyth (do.), Michael Summers (do.). Sergeants Daniel Donovan (Spoleto), James Coyne (do.), Richard Wall (do.), Corporals Patrick Lucy and Michael Hyde (do.), Privates Richard Cahill (do.), Patrick O'Shea (do.), John Reardon (do.), James Connor (do.), Michael Stapleton (do.), Andrew Daly (do.). Sergeant-Major Thomas Parker (Castelfidardo), Sergeant John Kirwan (do.), Private Matthew MacKenna (do.), Private James Lynch (do.). Sergeant John White (Ancona), Corporal Dolan (do.), Private—Nolan (do.), Private Murphy, Sixth Company (do.), Private Patrick Nevin (do.), Private Andrew O'Beirne (do.), Private Peter Murphy (do.), Private John Byrne (do.), Private—Ryan (do.).

APPENDIX C.

OFFICERS OF THE BATTALION OF ST. PATRICK IN 1860.

(Further details about them will be found in the Table of dates, Appendix D.)

Commanding Officer.

MAJOR O'REILLY, Myles William (Spoleto), of Knock Abbey, Co. Louth. For his services in this campaign created Commendatore del Ordine Piano. Previous service as a captain in the militia; had been a student of military matters. A sketch of his life appears in the *Catholic Encyclopedia.*

Captains

COUNT RUSSELL OF KILLOUGH, Francis (Ancona).— For servicesin this campaign created Cavaliere del Ordine Piano. Had been in the Papal army for two years before 1860, and had won a decoration. Afterwards became C.O. of the Company of St. Patrick; when it came to an end he joined the Papal Carbineers and fought at Mentana. He has left an interesting book describing his ten years in the Papal service. (*Dix années aw service du Pape.)*

BARON GUTTEMBERG, Francis Ferdinand (Ancona).— By birth a Bavarian, but had been an officer in the Austrian army, and was deputed to convoy the recruits from Ireland to Italy. He was selected for this service owing to his knowledge of English—being married to an Englishwoman of good family—and om arriving in Italy decided to join the battalion.

O'CARROLL PATRICK (Ancona).—Had previously been in the 18th Royal Irish; described by de La Moricière as an efficient officer.

BLACKNEY JAMES (Perugia).—Wounded at Perugia; created Cavaliere del Ordine Piano. Described by Mr. Crean as "of good family in Co. Carlow. Had been an officer in the county militia, and on his return to Ireland settled down to the life of a country gentleman." Mr. Crean must have mistaken the county, as this officer is elsewhere described as "son of James Blackney, Esq., of Kilmullen, Co. Kildare," but "grandson of the late Walter Blackney, M.P., for Co. Carlow."

KIRWAN, Martin (Castelfidardo).—"A militia officer, son of a Mr. Kirwan, who was an R.M. in Ireland; a cousin of Lord Howth." I see in *The Standard* (Dublin) for March 16th, 1929, that he and his brother were "sons of a widowed lady of independent means living at Thurles." Both are said to have been painstaking and patriotic officers.

COPPINGER, John (Spoleto).—Wounded at Spoleto, and created Cavaliere del Ordine Piano. Afterwards entered the American army, in which he served with great distinction, rising to the rank of General in their standing army, and, according to Mr. Crean, being considered one of the ablest officers of his day in the U.S. service.

BOSCHAN, August (Spoleto).—Had been an officer in the Austrian army; a good officer and popular with the men.

O'MAHONY, Francis (Ancona).—For his services in this campaign created Cavaliere del Ordine Piano. Born in Co. Cork, and served in the Austrian Uhlans (cavalry). He had fought in the campaign of Magenta and Solferino in 1859, and was considered a very smart officer. After this campaign returned to the Austrian army, and was seriously wounded at Sadowa in 1866.

Lieutenants and Second Lieutenants.

HOWLEY, Edward John (Spoleto).—Son of Mr. Howley, J.P., and D.L. of Belleek Manor, Co. Sligo. Had resigned a commission in the 10th Hussars to serve in Italy. After his return to Ireland he did not live many years.

CAREY (Castelfidardo).—A militia officer.

TERNAN, Augustus (Ancona). No details.

GREENE (Ancona).—A Tipperary man according to *The Standard*.

KEILEY, Daniel (Ancona).—Had been a midshipman in the British navy. For services in this campaign created Cavaliere dell' Ordine di S. Gregorio. Afterwards rose to command a cavalry brigade in the northern army during the American Civil War.

KIRWAN, W. (Spoleto).—Brother of Martin Kirwan abovementioned was a militia officer; only 17 years old. (*Standard* for March 16th, 1929.)

LUTHER, Martin L. (Perugia).—"Son of Mr. Luther of Clonmel, and nephew of the late Charles Bianconi," Mr. Crean told me. According to one informant he afterwards served as a captain during the American Civil War. Won the universal admiration of his men by his coolness under fire at Perugia.

CREAN, Michael T. (Spoleto).—His services during the campaign (wounded in defense of the gate at Spoleto, etc.) were rewarded with a Knighthood of Pius IX, an exceptional decoration for a subaltern. After his return to Ireland he was for many years a well-known barrister, and finally accepted an important post under the government (Legal Assistant Commissioner to the Land Commission Court).

KEOGH, Myles Walter (Ancona).—Afterwards rose to be a Colonel of American Cavalry, and A.D.C to General Stoneman. He was killed in the celebrated fight at Custer's Point, where the American troops were ambushed by Red Indians.

BUTLER, James H.—Gazetted August 4th, 1860. No

further details.

DOYLE, W. J.—Gazetted August 4th, 1860. No further details.

CRONIN (Spoleto).—For services during this campaign (notably his sharpshooting during the Spoleto engagement) created Cavaliere dell' Ordine di S. Gregorio. It is said that he afterwards emigrated to America, and served as an infantry captain during the Civil War.

DE LA HOYDE, Albert (Ancona).—Of ancient Catholic family living in Dublin; joined as 2nd Lieutenant. Being a keen soldier, Irishman and defender of the Pope, he afterwards enlisted in the Company of St. Patrick; rose to be a 2nd Lieutenant, and when it came to an end in 1862 passed on into the Papal Zouaves. In that celebrated corps he served with considerable distinction. He was promoted captain soon after the battle of Mentana, for gallant conduct in that engagement. In 1870 he was at the most exposed point outside the Porta Pia during the short defense of Rome. For him the fall of the Temporal Power wrecked a promising career. After the disbanding of the Papal Zouaves he was for many years in charge of the mails between London and Brindisi, being an accomplished linguist.

STAFFORD, William (Spoleto).—For services in this campaign created Cavaliere dell' Ordine di S. Gregorio, and then joined the Company of St. Patrick. I believe that some years later he rose to the rank of captain in the American infantry, apparently during the Civil War.

D'ARCY, James (Castelfidardo).—For gallantry at the battle of Castelfidardo he was created Cavaliere di S. Gregorio. Afterwards was an officer in the Company of St. Patrick, and then rose to be a captain in the Papal Zouaves. Fought at Mentana, served on the staff of the Emperor Maximilian in Mexico until the Emperor's death.

DUNNE (Ancona).—Gazetted 2nd Lieutenant, August

20th, 1860; but it seems to be probable that he fought at Castelfidardo with the Franco-Belgians.

THE MACSWINEYS (Ancona and Castelfidardo).—There were two officers of this name; apparently they were brothers. As far as I have been able to discover the eldest served in Ancona as a subaltern. He is described as "a member of a good family living at Carrigrohane Castle, Co. Cork."

The younger, according to *The Standard* of March 2nd, 1929, served first at Ancona, as a private. Wanting to obtain a commission he was advised by de La Moricière to enlist for six months in the Franco-Belgians in order to get a thorough training. He did so, together with Woodward and Dunne; at all events they are all classed together in a letter of de La Moricière (August 15th, A.S.R. Com.-gen. tr. pont. 3154), in which he describes them as being "neither privates nor officers," and on August 20th they were all gazetted 2nd Lieutenants. MacSwiney fought—and probably the other two with him—at Castelfidardo, among the Franco-Belgians, was wounded, and won the Order of St. Gregory. He is mentioned in the first quarterly return for 1861 of the Company of St. Patrick as a captain who had been wounded and a prisoner at Castelfidardo. No doubt he and his comrades were classified with the Irish Company (No. 4); they were certainly not mentioned among the Franco-Belgians on the list of prisoners.

WOODWARD.—Gazetted 2nd Lieutenant, August 20th, 1860; the son of a Protestant clergyman who became a Catholic. According to *The Standard* of March 2nd, 1929, he went with MacSwiney into the Franco-Belgians.

O'CONNELL, Michael (Ancona).—For services in the year 1861 created Cavaliere di S. Gregorio. He was by then an officer in the Papal Zouaves, and later rose to be a captain. He fought at Mentana.

WALSH, Joseph (Ancona).—Afterwards joined the Company of St. Patrick; retired, May 25th, 1861. Later is said to have served in the American Civil War.

O'KEEFFE, Joseph (Ancona).—A nephew of the Catholic Bishop of Cork.

LYNCH (Spoleto).—For services in this campaign created Cavaliere dell' Ordine di S. Gregorio.[268] Came of a military family, having a brother who was a V.C. Afterwards served as a private in the Papal Zouaves, where he is said to have shown himself a "gallant soldier."

LAWLESS, Alfred E. (Spoleto).—Died at Spoleto before the fighting began.

BARRY.[269]

O'FLYNN, Philip (Spoleto).—Lieutenant Assistant Surgeon. Afterwards joined the Company of St. Patrick, and when it came to an end passed into the Papal Zouaves, in which corps he was a well-known figure.

THE REV. EDWARD MACLOUGHLIN (Spoleto).—Chaplain (Firstclass). Volunteered, I believe, at the age of forty-five. Afterwards Chaplain to the Company of St. Patrick, and when it came to an end passed into the Papal Zouaves. (October 1st, 1862.)

Sergeant Majors.

MULHALL (Spoleto).—For his services in this campaign created Cavaliere dell' Ordine S. Silvestro. Formerly in an English Lancer Regiment. Afterwards served in the northern army during the American Civil War, distinguished himself at the engagement at Skinner's Farm. Rose to be a Colonel.

LLOYD (Spoleto).—His services were rewarded with a commission as 2nd Lieut., and the Cross of St. Sylvester.

---

268   About this officer I have been unable to find any details in 1860 except that he was gazetted Second Lieutenant on August 20th. 1860. and that on January 17th. 1861, he was decorated with the Ordine Piano, which was not very often given to subalterns. This may have been a reward for further services after the Campaign of 1860. He was by then an officer in the Papal Zouaves.

269   Of this officer I can find no records, and do not know whether his name has been inserted in the list by accident. It did not appear on my original list of officers.

Sergeant Major (or Sergeant).

GLEESON (Ancona).—Born in Co. Tipperary. He afterwards rose to be a Colonel during the American Civil War.
Heeney.
WHITE, Nicholas G.

On some lists I have seen the name of 2nd Lieutenant HEENY or Heany, but have not been able to find any trace of his being gazetted, or any details about him.

Also the name of Nicholas G. White, M.D., Assistant Surgeon, about whom I have found no details during the campaign. He is no doubt the same Dr. White who did such splendid work in collecting the wounded after the war was over.

APPENDIX D.

TABLE OF DATA FROM THE STATE ARCHIVES OF ITALY.

(Only about three quarters of the Officers are mentioned herin.)

The following is a table of facts and dates drawn exclusively from the Royal Archives in Rome. It is, I venture to believe, absolutely reliable, for apart from my own search, it is confirmed by that of the Marquis MacSwiney of Mashanaglass, M.R.I.A., who had made a close and systematic study of the subject, but then, finding that I had been engaged upon it for years, most unselfishly presented me with the results of his labors. Owing to this great act of generosity it has been possible to doublecheck the data; and to obtain an incontrovertible basis upon which to work.

In order to understand the table it must be remembered that the Irish battalion (the *Battalion* of St. Patrick) only remained in existence from June 12th, 1860, until the end of September of that year. But after the war was over, the Roman authorities decided to form an Irish *company*) the Company of St. Patrick. The Papal provinces being lost, the army was naturally re-organized on a greatly reduced scale; but as an acknowledgment of the services of the battalion the authorities arranged that its memory should be preserved in an Irish company of the same name.

Of the 35 or 36 Irish officers, some thirteen either joined or attempted to join the Irish company. It was, therefore, impossible to give them all commissions, and a waiting, list was instituted. But the privates did not return in anything like the same proportion. In Ireland it was felt that the danger to the Pope was now past; that his provinces were irretrievably lost, but he himself was under French protection. Those who re-enlisted were, in most instances, the men who had done best in the fighting in 1860, and even these enthusiasts did not all return. The company never numbered more than 46, and gradually dwindled down to a strength of only 22, before its end: its formation had been decreed on Nov. 8th, 1860, and it was finally disbanded on September 30th, 1862.

This short explanation seemed necessary for an understanding of the table.

In case of any reader wishing to consult the Archives for himself, the following are the references for which to ask:—
Regio Archivio di Stato, Roma.

1.—Archivio del Ministero delle Armi; Busta 1169.
Ordini del Giorno dal 4 Gennaio 1859 al 30 Luglio 1860.
Ordini del Giorno dal 10 Agosto al 31 Dicembre 1860.
Ordini del Giorno dal 3 Gennajo al 30 Dicembre 1861.

2.—Archivio del Ministero delle Armi; Busta 1171.
Ordini del Giorno del 9 Gennajo al 31 Dicembre 1862.
Ordini del Giorno dal 3 Gennajo al 10 Dicembre, 1863.

3.—Archivio del Ministero delle Armi; Busta 1861 (fogli di pagamento della Compa di S.P.)

4.—Compagnia di S. Patrizio—Matricola. Uffiziali.[270]

---

[270] I might add that it has been, incidentally, possible to verify many of the details related to me by veterans. But in this table nothing is included except the facts and figures recorded in the Archives.

NAME AND RANK	BIRTH		PARENTS	DOMINCILE
	DATE	PLACE		
Major: O'REILLY (Myles William)				Knock Abbey, Co. Louth
Captains: COUNT RUSSEL OF KILLOUGH (Francis)	1836. April 1st.	Vernon, Eure, France	Count Thomas Russell of Killough. Comtesse Ferdinande de Flammerant.	Pau, Basses Pyrénées, France
BARON GUTTEMBERG (Francis Ferdinand)				
O'MAHONY (Francis)				
O'CARROL (Patrick)				

Appendices

FOUGHT AT	DATE OF GAZETTE	SOME DETAILS OF SUBSEQUENT CAREER	Actions, Decorations, Etc., Up to the year 1862 I do not mention the medal Pro Petri Sede, decreed Dec. 8th, 1860, and carrying a year's seniority. All those who fought were entitled to it.
Spoleto	1860. Jun. 28th, Major		
Ancona	1858. Mar. 30th. 2nd Lieut. in 1st Foreign Regiment (Swiss) in Papal service. 1860. Mar. 21st. 2nd Lieut. in Battalion of St. Patrick. Aug. 7th, Captain.	1860. Nov, 9th. Captain of Company of St. Patrick. 1862. Oct. 1st. On Company of St Patrick coming to an end he joined the Papal Zouaves. But in 1864 he was gazetted to the Papal Carbineers	Taking of Perugia, 1859. *Decoration:* Knight of S. Gregory Nov. 22nd, 1860. Knight of Pius IX.
Ancona	1860. Aug. 4th, Captain	Afterwards appears on the waiting list of officers for the Company of St. Patrick, but apparently never joined.	
Ancona	1860. Aug. 7th, Captain		
Ancona	1860. Aug. 7th, Captain		

NAME AND RANK	BIRTH		PARENTS	DOMINCILE
	DATE	PLACE		
BLACKNEY (James)				
KIRWAN (Martin)				
COPPINGER (John)				
BOSCHAN (August)				
Lieutenants and Second Lieutenants: HOWLEY (Edward John)				
TERNAN (Augustus)				

FOUGHT AT	DATE OF GAZETTE	SOME DETAILS OF SUBSEQUENT CAREER	Actions, Decorations, Etc., Up to the year 1862 I do not mention the medal Pro Petri Sede, decreed Dec. 8th, 1860, and carrying a year's seniority. All those who fought were entitled to it.
Perugia (No. 1 Co.)	1860. Aug. 31st, Captain	Afterwards appears on the waiting list of officers for the Company of St. Patrick, but apparently never joined.	
Castelfidardo (No. 4 Co.)	1860. Aug. 31st, Captain		
Spoleto (No. 2 Co.)	1860. Aug. 31st, Captain Dec. 1st, retired		
Spoleto (No. 2 Co.)	1860. Jun. 27th, Second Lieutenant. Aug. 31st, Captain		
Ancona	1860. Jun. 24th, Second Lieut. & Adj. Joined Company of St. Patrick. Retired May 25th, 1861		

NAME AND RANK	BIRTH		PARENTS	DOMINCILE
	DATE	PLACE		
MaCSWINEY				
KEILEY (Daniel Joseph)[1]	1832.	Neewtown-ville, Waterford	Thomas Keiley Kate O'Byrne	Newtonville, Waterford
KEOGH (Myles Walter)	1840. March 25th	Orchard, Carlow	John Keogh Margaret Blanche	Carlow
BUTLER (James)				
DOYLE (James)				
DELAHOYDE (Albert)				

[1] Entered in one Gazette as William

Appendices

FOUGHT AT	DATE OF GAZETTE	SOME DETAILS OF SUBSEQUENT CAREER	Actions, Decorations, Etc., Up to the year 1862 I do not mention the medal Pro Petri Sede, decreed Dec. 8th, 1860, and carrying a year's seniority. All those who fought were entitled to it.
Ancona (or Castelfidardo?)	1860. Aug. 20th, Second Lieutenant.	*V.* Note to the MacSwiney of Castelfidardo.	Knight of S. Gregory. This decoration was awarded to "Captain" MacSwiney.
Ancona	1860. Aug. 7th, Lieutenant	1860. Nov. 9th. Joined the Company of St. Patrick.	Knight of S. Gregory
Ancona	1860. Aug. 7th, Second Lieutenant	1862. Feb. 20th. Resigned	
Spoleto	1860. Aug. 4th, Second Lieutenant	1860. Nov. 9th. Joined the Company of St. Patrick.	
Spoleto	1860. Aug. 4th, Second Lieutenant	1862. Feb. 20th. Resigned	
Ancona		Enlisted in the Company of St. Patrick; rose to be Sergt. Major 1862. Oct. 4th. Passed into the Papal Zouaves as Second Lieutenant. Promoted Captain after Mentana.	Of this officer there are other mentions in the Gazette up to 1870

NAME AND RANK	BIRTH		PARENTS	DOMINCILE
	DATE	PLACE		
STAFFORD (William)	1841 April 7th	Dublin	William Stafford Ellen O'Sullivan	Dublin
D'Arcy (James)	1844. March 13th	London	James D'Arcy Margarita O'Flynn	London
DUNE				
WOODWARD				
MacSWINEY				
O'CONNELL (Michael)				
WALSH (Joseph)				

Appendixes

FOUGHT AT	DATE OF GAZETTE	SOME DETAILS OF SUBSEQUENT CAREER	Actions, Decorations, Etc., Up to the year 1862 I do not mention the medal Pro Petri Sede, decreed Dec. 8th, 1860, and carrying a year's seniority. All those who fought were entitled to it.
Spoleto	1860 Aug. 7th, Second Lieutenant	1860. Nov. 9th. Joined the Company of St. Patrick. 1862 Oct. 1st. On its dissolution passed into the Papal Zouaves	Knight of S. Gregory
Castelfidardo	1860 Aug. 7th, Second Lieutenant	1860. Nov. 9th. Second Lieutenant Company of St. Patrick, 1862. Oct. 1st. On Company of St. Patrick coming to an end, joined the Papal Zouaves. Promoted Captain	Knight of S. Gregory Of this officer there are other mentions in the Gazette up to 1870.
Ancona	1860. Aug. 20th, Second Lieutenant		
Castelfidardo or Ancona	1860. Aug. 20th, Second Lieutenant		
Castelfidardo (with Franco-Belges)		For MacSwiney V. App. C	
Ancona	1860. Aug. 20th, Second Lieutenant		
Ancona	Second Lieutenant		

NAME AND RANK	BIRTH		PARENTS	DOMINCILE
	DATE	PLACE		
O'KEEFE (Joseph)				
O'FLYNN (Philip) (*Surgeon*)	1831 July 27th	Clonmel	Patrick O'Flynn Margaret Kennedy	Waterford
MACLOUGHLIN	1815?	Dublin	Edward MacLoughin Maria Murphy	Dublin

In addition to the above there were at least four foreign officers or N.C.O's transferred to the Irish battalion:—

RUPP (Andrea), Swiss, for Instruction;
SCHMIDLIN (Carlo), Swiss, Paymaster;
LUZI (Giovanni), later replaced by Meinhold (Ottone) for Clothing.

These non-combatant officers were given the rank of Lieutenant or Second Lieutenant: they no doubt figured among the prisoners taken at Spoleto. Besides these there were several drill-sergeants, and, no doubt, some privates for orderly-room, etc.
Captain MAZZOLI and 2nd Lieut. ANETHON were also named, but whether they ever joined is uncertain.

Appendices

FOUGHT AT	DATE OF GAZETTE	SOME DETAILS OF SUBSEQUENT CAREER	Actions, Decorations, Etc., Up to the year 1862 I do not mention the medal Pro Petri Sede, decreed Dec. 8th, 1860, and carrying a year's seniority. All those who fought were entitled to it.
Ancona	Second Lieutenant		
Spoleto	1860. Aug. 10th Lieutenant Assistant Surgeon	1860. Nov. 9th Joined the Company of St. Patrick. 1862. Oct. 1st. Passed on to the Papal Zouaves	
Spoleto	1860. April 30th. Chaplain. (First Class)	1860. Nov. 9th. Joined the Company of St. Patrick as Chaplain (First Class). 1862. Oct. 1st. Passed to the Battalion of Zouaves	

APPENDIX E.

THE NUMBER OF IRISH WHOU FOUGHT IN THIS CAMPAIGN.

The exact numbers of men in the four Irish companies trained at Spoleto are now impossible to discover. No. 1 Co. went to Perugia, Nos. 2 and 3 remained at Spoleto; No. 4 fought at Castelfidardo—except about 30 of its men who returned to Spoleto.

The Italian official history of the campaign quotes a document dated August 1st, 1860, which states the total number of Irish in Italy as 1108 of whom 450 were at Ancona, and 646 at Spoleto; 12 in Rome. But one cannot trust this total of 646 for the southern four companies, because the return is evidently inaccurate. It says that no changes had taken place since July 1st, whereas 118 Irish had gone home, and various batches of recruits had arrived and it does not seem to fit in with any of the other statements of strength.

The nominal strength of each company was 140, including officers. (Busta 3155. Fascicolo 46. Order of July 4th.)

The difficulty of finding out the *exact* strength of the companies lies in the fact that the officers nearly always give a round figure. Thus de La Moricière stated No. 1 Company at 130, O'Reilly stated No. 4 Company at 130, and gave 300 as the total number of Irish in Spoleto during the fighting. These are merely round figures. On the other hand, when Brignone says he took 345 Irish in Spoleto this figure seems to be too high to square with any of the others.

One can only give the following estimate: there are two calculations which do square with one another more or less.

(1.)—On July 24th Pimodan stated the total number of Irish at Spoleto at 682. After that date 163 returned home; but at the same time about 48 new recruits arrived in Italy. There would therefore have been about 567. To these must be added the 12 still in Rome. This would give a total of 579 in the four Spoleto companies.

(2.)—In Ireland a useful pamphlet (*The Irish Papal Brigade Vindicated*) was published shortly after the campaign to defend the Irish battalion against the Times and other papers, and containing the report of a speech by Major O'Reilly. It states the number as follows:—

At Spoleto (according to O'Reilly) ...	300 men.
At Perugia	143 ,,
At Castelfidardo	105 ,,
	548

As it stands this statement seems too low to be possible. But it squares fairly well with (1.) if we assume that it omitted the officers, as is not uncommon in such calculations. Thus:

At Spoleto: 300 privates and N.C.O's would be the nearest round figure; it was given by O'Reilly after he had left the State, and seems to be a trifle below the mark. There were two companies with nominal strength of 137 men each, plus the 30 men sent back from No. 1 Co. This gives a total of 304 rank and file, which was, no doubt, the idea in his mind. As a matter of fact one is rather inclined to accept a total of 312 given by a volunteer named Hogan, who fought in Spoleto. And besides these there were 15 officers.

At Perugia: 143 privates and N.C.O's. This No. 1 Company was reported as marching 140 strong, probably a round figure. De La Moricière had spoken of it as 130 strong, but when it became "the active service Co."; everyone wanted to be transferred to it so as to see fighting.

It had two officers while in Perugia.

At Castelfidario: 105. De La Moricifere had ordered this Company to be reduced to 100, but his round figure was liberally interpreted, as there was great keenness to go to the front. Its full strength was originally stated as "130," which no doubt is also a round figure, meaning 130-and-odd men. It could hardly have been so much below strength as 130 when the other companies were over-full. Besides these there were its three officers.

Thus the four southern companies total up to 580, which is very near to the (1) calculation of 579.

In Ancona: 450. Also a round figure. There seem to have been 456 Irishmen in Ancona, of whom 16 were officers.

Consequently, the probable totals of those who actually fought seem to be:—[271]

	Men.	Officers.	Total.
Spoleto[272]	312	15	327
Perugia	143	2	145
Castelfidardo	105[273]	3	108
Ancona	440	16	456
	1000	36	1036

The total given in the text (Chapter III) is "approximately 1040"; this allows for a few being in hospital at Spoleto. The

---

271   The following numbers are estimates and not necessarily exact.
272   Including the chaplain, doctor, and supernumeraries. The large number of officers at Spoleto is due to the fact that there were several young supernumeraries there under training—four, I think.
273   This seems to me to be a minimum figure.

only hospital return that I have been able to find was one for Ancona in the beginning of August; it showed that there were only four Irishmen on the sick list.

APPENDIX F.

THE IRISH CASUALTIES.

These are to a certain extent a matter of estimation, because owing to the smallness of the Irish units their officers did not write any reports except at Spoleto.

At Perugia we know that in Allman's party alone there were at least eight casualties; and, apart from them, Captain Blackney was wounded, and apparently Diamond and one of the sergeants as well. As the men were fighting both in the streets and in the fort it does not seem excessive to suppose that their casualties were, at the very least, 15 in all; perhaps 20.

For the two companies at Spoleto O'Reilly reported three killed and ten wounded; but Edward Dunne told me that there were others who did not report themselves so as not to be kept in Italy; he thought there would be "hardly 20 in all;" say 17 or 18.

At Ancona most of the veterans said there had been over a dozen casualties; on one day alone there were 10 or 12. O'Beirne I think died on the way home, and Skehan died in Cork—but there are no records. Say 12 or 14 casualties in all.

For Castelfidardo there is practically no information—except the statement by Mr. W. P. Ryan's friend that there had been "between 20 and 30 wounded," and the statement in the Italian official account of the campaign which says, "the casualties of the Irish company are not known, but they were certainly heavy." Say 20.

This gives a total estimate of about 65 to 70. Its comparative lightness is due to the fact that the four companies in Ancona had very few casualties, because the surrender of the place was brought about by the fleet on the very day before the closest bombardment of the main forts could begin; thus the 456 Irishmen there escaped with only about a dozen casualties. The southern half-battalion must have had about 55 or 56 men hit—nearly 10 percent.

APPENDIX G.

NOTES ON THE LOSSES AT SPOLETO.

The losses at Spoleto were surprisingly small on both sides. Major O'Reilly returned his at only 3 killed and 10 wounded. But, as a matter of fact, there were a few more. The slightly wounded men did not report themselves as such, from fear of being left in Italy when their comrades went home. Still,—so Edward Dunne told me—the total number of casualties was "hardly twenty."

On the Piedmontese side General Brignone's report has also, perhaps, a slight tendency to minimize matters. The capture of the place had proved a more serious matter than had been anticipated, and he felt probably that he had made a cardinal mistake in ordering an assault before the way had been fully prepared by artillery. The totals given in the official returns are 14 killed and 49 wounded; but since, then a monument has been raised to those who fell, giving their names, and stating them at 16 killed and 49 wounded. This total includes no officers, of whom we know that some were hit, and mentions no civilians.

Major O'Reilly estimated the Piedmontese losses at, probably, about four hundred, but he was evidently misled by the figures given him. He based his estimate, apparently, on two definite items: (1) after the fight he was informed, to his surprise, by General Brignone that there had been 100 wounded (not killed) on Monte Luco. This was manifestly a misunderstanding, probably due to difference of language. The Italian returns do not admit any casualties at all on Monte Luco: (2) O'Reilly also tells us that a Bersaglieri officer showed him the returns of his company, and that these were nine killed and 22 severely wounded.

As a matter of fact this latter instance was perfectly accurate. Some fifty years later the returns of the 35th Bersaglieri were published, and they give nine killed and 21 wounded, omitting as already stated, the officer. So that on this point O'Reilly was absolutely correct. But this company had suffered far more than any other, because it took part in the assault.

Arguing from these bases O'Reilly undoubtedly formed too high an estimate.

One should, however, add (3) that even he did not quite realize the ineffectiveness of the muskets. After firing all day he naturally supposed that they would have inflicted some losses on the enemy; but it seems that, apart from the assault, the Piedmontese losses must have been almost entirely due to the few rifles of the garrison.

BIBLIOGRAPHY.

BOOKS:

ALESSANDRINI, ALESSANDRO. *I fatti politici delle Marche, del I Gennaio, 1859, al epoca del plebiscito.* A detailed history of the Marches during 21 months—very anti-Papal. Valuable for its documents.

A.S.R.U. *Archivio storico risorgimento Umbro.* A historical quarterly for Umbria. Most interesting.

AZZI, G. DEGLI. *L'Insurrezione e le stragi di Perugia.* 1009. —*Per la liberazione di Perugia.* 1910. (A most interesting history of Perugia and Umbria during the year 1860. As. to the Irish battalion he repeats the *Nazione* accusation, but has since then expressed a far more favorable opinion. *V. Risorgimento italiano* (a historical review) for 1913, p. 863 note.

BECDELIÈVRE, COMTE DE. *Souvenirs de l'armée pontificale.* When describing this campaign he is naturally proud of his Franco-Belgians, but a little prone to depreciate the other battalions.

BIANCHI, NICOMEDE. *Storia documentata della Diplomatic Europea in Italia dall' anno 1814 all' anno 1861.* Turin, 1869. 5 vols.

BITTARD DES PORTES, RENÉ. *Histoire des Zouaves pontificaux.* Interesting; but it is a history of the Papal Zouaves (not of the Papal army), and consequently can only give a short notice to this campaign; in which, however, the Irish battalion is favorably mentioned.

CARANDINI, MARCHESE FEDERICO. *Manfredo Fanti; sua vita* (*Life of General Fanti*. 1872.)

CHARETTE, BARON DE. *Souvenirs du Regiment des Zouaves Pontificaux*. 1871. Gives a rather different version of O'Reilly's report from any that I have seen.

CHIALA, L. *Ricordi della vita di due Generali Italiani*. F. Brignone and Giovanni Durando. Articles from the *Opinione* by this well-known writer.

COLLEVILLE, COMTE DE. *Un crime du Second Empire*. A pro Papal account of the campaign.

CONYNGHAM, CAPTAIN D.P., A.D.C. *The Irish Brigade and its Campaigns*. A short and interesting account of Meagher's. Brigade, in which we find mentioned various ex-volunteers. of our Papal battalion.

CORSI. *Venticinque anni in Italia*. Corsi worked for the Italian, H.Q. On this campaign he was a captain on the staff of di Savoiroux.

CORVETTO, MAJOR. *La campagna di guerra nell' Umbria, e nelle Marche*. Semi-official. A valuable military account.

CREAN, M. T., in the *Seven Hills Magazine* for March, 1908. He distinguished himself at Spoleto.

FINALI, GASPARE. *Le Marche.*

FLORNOY, EUGENE. *La Moricière*. 1912. One of the series of *Les grands hommes de l'Eglise.*

FUMI. *Orvieto.*

KELLER. *Le General de La Moricière*. Life of de La Moricière.

LECOMTE. *L'Italie en 1860*. He was a Major on the H.Q. Staff of Switzerland.

MATHUISIEULX, H. M. DE. *Les Zouaves Pontificaux.*

O'CLERY, THE. *The Making of Italy. 1892.*

O'REILLY, MAJOR M. W. *Diary* (unpublished) of the greatest interest.

ORERO. *Da Pesaro a Messina.* (*Memoirs*) 1905. General Orero was A.D.C. to Cialdini, and gives a vivid picture of his campaign.

PERSANO. *Diario privato-politico-militare.* 1870. Persano's diary for 1860—I.

PIMODAN, GABRIEL DE. *Vie du General de Pimodan.* A life of Pimodan by his son.

POLI, VICOMTE OSCAR DE. *Les soldats du Pape.* 1868. *De Paris a Castelfidardo. Paris,* 1867. *Souvenirs du Bataillon des Zouaves pontificaux.* 1861.

PRAMPERO. *La Brigata Regina. Da Bologna per Castelfidardo a Gaeta. 1910.* Short but interesting.

QUATREBARBES, DE. *Souvenirs d'Ancone.* The work of a charming, absurdly prejudiced and gallant old legitimist who at the age of 57 left his home and his invalid wife, to fight for Church and King. He was appointed governor of Ancona.

REVEL, GENOVA DI. *Da Ancona a Napoli; miei ricordi.* He was Fanti's staff officer for Artillery. *Umbria ed Aspromonte.* A short account of Masi's column.

ROCCA DELLA. *Autobiografia di un veterano.* Autobiography of General Della Rocca.

ROMANO, UN. Castelfidardo e Ancona.

ROMAIN, UN. *La Bataille de Castelfidardo.* This work has sometimes been attributed to de la Moricière's inspiration, but the Italian version is the best on account of the notes of the translator, who is said to have been a Papal gendarme officer.

RUSSELL OF KILLOUGH, COUNT FRANK. *Dix anniés au service pontifical.* Very interesting; and valuable for our subject, as the author was in command of one of the Irish companies in Ancona.

RUSTOW. *Erinnerung aus dem italienischen Feldzuge von*

1860. Good.

SALICETO, ALFONSO VISCONTI DI. *Da Livorno a Napoli, 1860.* 1907. Reminiscences of a Subaltern with Della Rocca's Force.

SANDONNINI, T. *In Memoria di Enrico Cialdini.* 1911. A life of Cialdini.

SANTORELLI. *La Presa di Spoleto.* 1875. The writer was an eye-witness of the fighting at Spoleto; his first little book, issued in 1875, is rather better than the edition of 1910. He was a boy in 1860, and well remembered Major O'Reilly and the Irish.

SEGUR, MARQUIS DE. *Les Martyrs de Castelfidardo.* Paris, 1891.

STATO MAGGIORE (Italian War Office.) Official works produced by the Ufficio Storico (historical office). These works are extraordinarily thorough.

BARBARICH. *La Battaglia di Castelfidardo.* 1903. A good description of the battle.

CESARI, COL. CESARE. *I cacciatori del Tevere.* A thorough and impartial account of all Masi's operations around Orvieto and afterwards.

VIGEVANO, COL. A. H.—*La fine dell' esercito pontificio.* 1920. An exhaustive description of the Papal army. *La campagna delle Marche e dell' Umbria.* 1923. A masterly description of the campaign of Castelfidardo. It leaves nothing more to be said.

SULLIVAN, A. M. *New Ireland.* 1877.

TOURNON, LE COMTE DE. *Les volontaires pontificaux à cheval.*

TREVELYAN, GEORGE MACAULAY. *Garibaldi and the Making of Italy.* 1911. Gives an impartial sketch of the campaign of Castelfidardo.

VEUILLOT, LOUIS. *Les Piémontais dans les Etats de l'Eglise.* 1861.—A pro-Papal history of the campaign.

REVIEWS:

The *Risorgimento Italiano, Rivista d'Italia, Rivista militare, Memorie storiche militare*, and many other publications of this nature.

A FEW GENERAL HISTORIES OF THE PERIOD:

(Cantu is Papal. Johnson is Papal, but very impartial. Pelczar is clerical, and the others are all in favor of United Italy; the best Italian historian of the Risorgimento is, I think, Tivaroni).

CANTU. *Cronistoria*. Tome III.

CELLAI. *Fasti Militari*. Vol. IV.

COPPI. Annali. XV. 1860.

JOHNSON, THE REV. HUMPHREY. *The Papacy and the Kingdom of Italy*. A handbook; an excellent summary.

MASI. *Il Risorgimento italiano*. 2 vols.

ORSI. *Histoire de l'Italie modern*.

PELCZAR. *Pio IX e il suo pontificato*. A good life of Pius IX by a Polish bishop. It has been translated into Italian.

ROSI. *Storia contemporanea d'Italia*. 1922.

TIVARONI. *L'Italia degli Italiani*. II. 1859—1866.

OFFICIAL REPORTS:

De la Moricière's Report. Published in book form.

Fanti's Report. Published in book form.

Cialdini's Report. (Advance on Castelfidardo side.) Italian War Office (Ufficio Storico).

Della Rocca's Report. (Advance on Perugia.) Italian War Office (Ufficio Storico).

Cadorna's Report. (Advance on Gubbio and Ancona). Italian War Office (Ufficio Storico).

De Sonnaz' Report. (Perugia, and after.) Italian War Office

(Ufficio Storico).

Di Savoiroux' Report. (Perugia, and after.) Italian War Office (Ufficio Storico).

General Schmidt's Report. (Perugia.) Vatican Archives.

General Schmidt's Report. (Confidential Report.) Archives. Vatican

Colonel Lazzarini's Report. (Perugia.) Vatican Archives.

Brignone's Report. (Spoleto.) Italian War Office (Ufficio Storico).

O'Reilly's Report. (Spoleto.) Italian War Office (Ufficio Storico).

Artillery Report. Operazioni dell' Artiglieria negli Assedii di Gaeta e Messina (includes siege of Ancona).

Engineer's Report. (General Menabrea.) Il genio nella campagna d'Ancona, etc. 1864.

Less important reports, such as Becdelièvre's, Blumensthil's, etc., are, for the most part to be found in the Italian War Office; but some are in the State Archives.

The State Archives of Italy are referred to by means of the following abbreviations.

A.S.R. Com. Gen. tr. pont. (Archivio di Stato Roma. Commando generale delle truppe pontificie.)

A.S.R. M.A. Aff. Spec. (Archivio di Stato Roma. Ministero delle armi, Affari Special.)

A.S.R. M.A. Aff. Ris. (Archivio di Stato Roma. Ministero delle armi. Affari riservati.)

A.S.R. Comp. S. Patrizio. (Archivio di Stato Roma. Compagnia di S. Patrizio.)

For the Archives of the Italian War Office, the following abbreviations are used:—

Uff. Stor. I. (Archivio dell' ufficio storico volume o cartella I.) For the Vatican Archives no abbreviation is used in this book. The Communal records at Ancona, Perugia, Spoleto and elsewhere have also been consulted. Documents from the British Public Record Office in London come under the reference F.O. Rome. Vols. 81 and 82.

NEWSPAPERS:

The *Nazione,* the *Monitors, Toscano,* the *Giornale di Roma,* the *Opinione* and many other Italian, French, English and Irish papers. The most useful is the *Nation.*

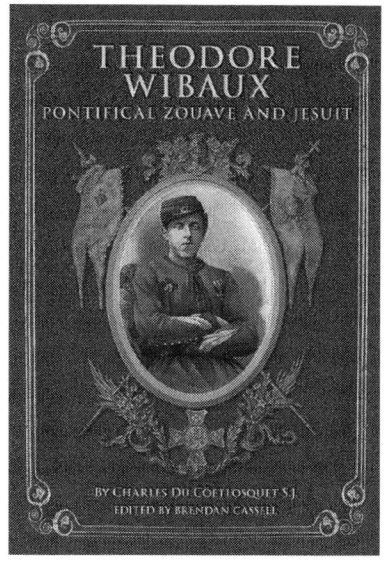

Check out Papal Zouave International's other republished books at **PapalZouave.com**

Made in the USA
Coppell, TX
06 October 2025

60902609R00176